307

biodiversity

a beginner's guide

From anarchism to artificial intelligence and genetics to global terrorism, Beginner's Guides equip readers with the tools to fully understand the most challenging and important debates of our age. Written by experts in a clear and accessible style, books in this series are substantial enough to be thorough but compact enough to be read by anyone wanting to know more about the world they live in.

anarchism
ruth kinna

anti-capitalism
simon tormey

artificial intelligence
blay whitby

biodiversity
john spicer

bioterror & biowarfare
malcolm dando

the brain
a. al-chalabi, m.r. turner & r.s. delamont

criminal psychology
ray bull *et al.*

democracy
david beetham

energy
vaclav smil

evolution
burton s. guttman

evolutionary psychology
robin dunbar, louise barrett & john lycett

genetics
a. griffiths, b. guttman, d. suzuki & t. cullis

global terrorism
leonard weinberg

NATO
jennifer medcalf

the palestine–israeli conflict
dan cohn-sherbok & dawoud el-alami

philosophy of mind
edward fesar

postmodernism
kevin hart

quantum physics
alastair i. m. rae

religion
martin forward

the small arms trade
m. schroeder, r. stohl & d.m. smith

Forthcoming:

astrobiology
lewis dartnell

asylum
pamela goldberg

capitalism
andrew kilmister, gary browning

cloning
aaron levine

conspiracy theories
alasdair spark

extrasolar planets
ian stevens

fair trade
jacqueline decarlo

forensic science
jay siegel

galaxies
joanne baker

human rights
david beetham

immigration
liza schuster

the irish conflict
anthony mcintyre, david adams

mafia
james finckenauer

racism
alana lentin

radical philosophy
andrew collier

time
amarendra swarup

volcanoes
rosaly lopes

biodiversity

a beginner's guide

john i. spicer

ONEWORLD
OXFORD

biodiversity: a beginner's guide

Published by Oneworld Publications 2006

Copyright © John Spicer 2006

All rights reserved
Copyright under Berne Convention
A CIP record for this title is available
from the British Library

ISBN-13: 978–1–85168–471–7
ISBN-10: 1–85168–471–9

Typeset by Jayvee, Trivandrum, India
Cover design by Two Associates
Printed and bound in India for Imprint Digital

Oneworld Publications
185 Banbury Road
Oxford OX2 7AR
England
www.oneworld-publications.com

Learn more about Oneworld. Join our mailing list to find out about our latest titles and special offers at:

www.oneworld-publications.com/newsletter.htm

We take a tiny colony of soft corals from a rock in a little water world.
And that isn't terribly important to the tide pool.

Fifty miles away the Japanese shrimp boats are dredging with overlapping scoops, bringing up tons of shrimps, rapidly destroying the species so that it may never come back, and with the species destroying the ecological balance of the whole region.
That isn't very important in the world.

And thousands of miles away the great bombs are falling and the stars are not moved thereby.

None of it is important
or all of it is.

John Steinbeck, *Log from the Sea of Cortez*

For Fiona, my fair one

contents

preface

It is said that books are best written in community. Over the past fifteen years I have been extraordinarily fortunate in the scientists I have worked with or for. They have made a lasting impression on what I know and believe about biodiversity, and this book would not have been written without their input in so many ways. I owe so much to Lorraine Maltby, Phil Warren, Dave Morritt and Kevin Gaston for providing such a stimulating and exciting environment in which to work and think when I was at the University of Sheffield, and Kevin in particular as he opened my eyes to the notion that biodiversity was a serious science. I feel privileged to have spent so much of my time at Sheffield discussing, investigating and writing with Kevin, and I thank him for allowing me to use the same broad outline for introducing novices to biodiversity that we came up with in the *Rising Sun* so many years ago.

I also thank my present colleagues, the members of the Marine Biology and Ecology Research Centre here at Plymouth – Rikka, Simon, Dave, the 'Bish', Kath, Mark, Martin, Pete, Andy, Paul, Miguel, Mal, Kerry, Jason and Steve – for their friendship and for making going into work on a Monday morning something to look forward to; all of the postgraduate students, postdoctoral fellows and academic staff with whom I have had the honour of working and so adding just a little to our knowledge of what biodiversity is and how it works – Sally Marsh, Kirsten Richardson, Jeanette Sanders, David Johns, Tony Hawkins, Steve Widdicombe, Nick Hardman-Mountford, Mike Kendall, Nikki Dawdry, Jenny

Smirthwaite, Kate Arnold, Lucy Dando, Emily Hodgson, Anne Masson, Sanna Ericksson, Sussie Baden-Pihl, Jalle Strömberg, Peter Tiselius, Jenny Cowling, Jason Weeks, Andy Rees, Mona Mabrouk El-Gamal, the inimitable Dave Morritt, Alan Taylor, Andy Hill, Stuart Anderson, Warren Burggren, Roy Weber, Brian McMahon, Peter Duncan, Katherine Turner, Alistaír Edwards, Peter Spencer Davies, Maria Thomasson, Bjent Liljebladh, Paul Bradley, Angela Raffo, Hayley Miles, Ula Janus, Hugh Tabel, Tim Blackburn and the inspirational Geoff Moore who, as well as co-supervising my doctorate, first opened up to me the wonder and science which characterizes the best of biodiversity as an academic subject. I am grateful to Roger Byrne and Mick Uttley, two of the sharpest minds I've ever encountered, for their detailed feedback on the manuscript and Marsha Filion of Oneworld, Terry Williams and Richard Wallace who also made some helpful comments on the manuscript. Suffice to say none of the above are responsible for any errors, omissions, transgressions or biases that remain. Although too numerous to mention by name, I certainly owe a large debt to all my undergraduate students who have taught me so much as they caught on to how exciting and threatened our biodiversity is.

I seem to have gone through a fair number of editors at Oneworld, but my thanks are no less heartfelt to Victoria Roddam from that initial meeting in a coffee shop in Bath (of all places) where the whole project kicked off, Mark Hopwood who had to badger me for so long that (unrelatedly) he gave up work and went back to becoming a student of philosophy, and finally an even more long-suffering Mike Harpley and Marsha Filion. All have been marvellous in their patience and help.

Finally, thanks to Fiona, my wife, and my children, Ellie, Ethan and Ben, for being so understanding and supportive. Ben provided the art work for the book, more than making up for my lack of ability in that area, and for that I am grateful. And it is to Fiona, my fair one, that I dedicate this book. As she well knows, none of this would have happened without her.

John Spicer, *2006*

chapter one

the wood among which the trees are found

But of course, for those of us who understand life, we could not care less about figures.

The Little Prince, Antoine de Saint Exupéry

biodiversity – what was that again?

Like it or not, 'biodiversity' is one of the big buzzwords of our time. You can hear it on the radio and in conversation, on TV and in films. Talk of the 'biodiversity debate', our 'biodiversity crisis', 'threats to biodiversity' (from climate change, a new motorway or GM crops), and 'conserving biodiversity' (by setting up nature reserves, stopping a new housing estate being built, paying to adopt a lion, panda or dolphin, letting your garden run wild) is everywhere. It's a word frequently found on the lips of politicians, ecowarriors, broadcasters, business people, university students, your friends and acquaintances down at the pub or cafe, conservationists, and even schoolchildren. And yet trying to pin down exactly what all of these different types of people mean by biodiversity is difficult. It seems to mean different things to different people. So we have a subject that many of us would agree is essential to know something about, even to get to grips with, but one for which few of us have a clear definition.

Fortunately, if you know where to look you can find some definitions for the word – the problem is, if you look quite hard you can find more than eighty different definitions. There is, however, one that has gained international currency, signed up to by the 150 nations that put together the Convention on Biological Diversity at Rio de Janeiro, Brazil in 1992.[www#1] Here biodiversity was defined as 'the variability among living organisms from all sources including [among other things] terrestrial, marine and other aquatic ecosystems and the ecological complexes of which they are a part ... [including] diversity within species, between species and of ecosystems'. In short, biodiversity is the variety of life – in all its different forms and relationships. This sounds quite satisfying until we ask the question that should let us know if the study of biodiversity is a science: how do we measure biodiversity? This is not so straightforward. And yet it goes to the very heart of what we mean when we talk about biodiversity or the biodiversity of a particular area, country or region. A simple example will help us see where the difficulties lie.

life – in abundance

I live a few minutes' walk from Wembury Bay (Figure 1). It's an incredibly beautiful bay on the south-west coast of Devon, just a few miles east of the city of Plymouth.[www#2] It was one of the first voluntary marine reserves in the UK. This was in recognition of the tremendous variety of life found there; the biodiversity of Wembury Bay is certainly impressive. Between the tides, on this shore alone, nearly all the different major 'types' or 'designs' of the things that characterize life on earth can be found. There are fish, sea urchins, crabs, worms, sea squirts, sponges, sea anemones, seaweeds, sand flies, bootlace worms, limpets, periwinkles, sea mats and lichens in abundance. Even the rocks in the vicinity of the bay harbour the remains of sea creatures which lived here

Figure 1 *Wembury Bay, Devon, UK*

more than 400 million years ago. There are armoured fish and crinoids (upturned starfish with stalks), quite different from today's marine animals but also eerily familiar. As John Steinbeck wrote of a similar, but totally different, seashore, 'Here we have life, and life in abundance.'

And there is the unseen microworld of Wembury Bay. A world within a world, ubiquitous on the surfaces of rock, animals, plants, refuse, between and around single sand grains, enclosed in a water droplet, and even living inside the bodies of the animals, plants and other microforms of life on the beach. Within this microworld viruses and bacteria abound, as do many minibeasts and plants that are so unfamiliar they have no common name or title. They range from nondescript, wormlike forms through to exquisitely formed, delicate and bizarre water bears living beneath the rims of barnacle shells.

Let us return to where we started: how do we measure biodiversity, and, in this case, the biodiversity of Wembury Bay? How should we go about answering this question?

The simplest way might be to count how many different types of things are there. This is no mean task, and could potentially take not just weeks and months but even years and tens of years, even for such a small area – and that's leaving out all of the land animals, plants and microbes that form part of the larger landscape that is the bay. It is just about conceivable that for most of the largish marine animals we could put this list together by drawing on an amazing book that compiles scientific records from Plymouth (and Wembury in particular) stretching back into the nineteenth century, the *Plymouth marine fauna.*[www#3] The seaweeds too we could probably get from a number of published scientific sources. However, for many microscopic animals, plants, fungi, bacteria and viruses our current information is sketchy at best. It's not just the process of finding them that is problematic either. Many of these little specks of life are yet to be described, let alone the number of different types counted.

Even if it were possible to count the numbers of different types of everything, would such a list really be the measure of the biodiversity of Wembury Bay? Well, perhaps. But it ignores the fact that there are rarely equal numbers of everything. Some creatures are extremely numerous and ubiquitous, others are rare or only occur in particular, sometimes very localized, areas. Periwinkles are absolutely everywhere, sometimes in large piles two or three animals deep. Sea cucumbers such as the white cotton spinner can be found under rocks at most times of the year but you really do have to search. Surely biodiversity must encompass not just differences but the actual numbers of different things present? But even that is not all.

Up until now all the differences we've considered have been determined by what the creatures look like and how that differs from how others look. It does not take into account that other differences may be equally, or even more, important; an example would be differences in the way individual types of creature 'work', i.e. how they acquire energy and what they do with that energy to maintain themselves and what they actually contribute (if anything) to the working of the ecosystem to which they belong. (We've already come across the term 'ecosystem' when we looked at the definition of biodiversity but I didn't say what one was. An ecosystem is a dynamic complex of plant, animal and microbial communities and their non-living environment interacting as a functioning unit. It can be small, e.g. a rock pool, or large like the Arizona desert, and anything in between.) We know that at Wembury limpets and sea hares, two different types of animal, often 'do' the same sort of thing – they do what cows do on land, they graze. And then there are the interrelationships between the different species – predators (the eaters) and prey (the eaten), for example – or the multitude of ways that species and even groups of species influence other species or groups.

If you've been following all this you may find yourself at a mental crossroads. This is a well-trodden road. It is a place where many scientists, philosophers and theologians find themselves periodically, and it is a place that we will return to time and again in what follows. You can go down the 'Oh, but the world's a complicated place and we'll never get to grips with it' road, which leads to a comfy armchair, subdued lighting, a stiff drink, and an abandoning of intellectual pursuit and its partner hope. Or you can opt for 'OK, it is complicated and I may never find the truth, but I'll settle for a little less if it keeps me from stalling and keeps me walking down this particular road'. This said, in the face of such complexity it is fair to say that there is no one way of measuring or quantifying 'biodiversity'. We cannot measure the biodiversity of Wembury Bay, or any other bay, or of the oceans, or of the earth for that matter. We can talk and think about the notion of biodiversity, but we cannot measure it – we can only measure *selected aspects* of it. Don't despair, though. It may not be ideal, but even that is a start.

directions

To put together a beginner's guide to biodiversity, based on current scientific knowledge and understanding, much of our time will be spent on looking at measures of biodiversity and how those measures change in time and space. Some will be better than others. In many cases, though, we will find that the measure has been decided for us. Scientists often have to rely on the total number of species, the species richness, in a given geographical area just because that is the only information available. Much work has gone into producing alternative measures. But given the data we already have in scientific literature and museums, the relative ease of putting together inventories of different types of creature, particularly for very large areas, and the fact that it often encompasses numerous aspects of biodiversity, species richness is not a bad measure. So much of what follows will use biodiversity and species richness almost as interchangeable terms: but not all the time.

There is no one way to write a beginner's guide to biodiversity. It could take the form of an exhortation to save the planet; it could be encyclopaedic, cataloguing the types of living creature and the places they live; it could centre on how to preserve biodiversity; or it could combine aspects of all three. So what will be the approach of this beginner's guide? Many theologians, philosophers and educators believe that the only way you can ever say anything general and all-embracing, is by starting with something tangible, specific, familiar. In the nineteenth century Thomas Huxley, for example, used the crayfish for the title and subject of a book he wrote to introduce interested readers to the study of zoology. In that same tradition, throughout this book I use Wembury Bay, and other aspects of my own experience, as a way into some of the big biodiversity issues and patterns. In that respect this is a very personal book.

What will be the key features of this beginner's guide? In the next chapter we ask the questions how many species are there currently on the earth, and how are they distributed between the different large groupings of organisms we currently recognize. What are these large groupings and how have we ended up with them? This will involve trying to determine what makes a species a species anyway. We will spend some time looking at numbers of species found in a particular area as an indication of, indeed as a surrogate for, the biodiversity of that region. In chapter 3 we will see that biodiversity is not distributed evenly across the earth's surface. There are hotspots and there are coldspots. We will look at the current patterns of biodiversity (or at least measures of biodiversity), in particular how the number of organisms varies with latitude, altitude and depth. That should take us neatly on to the fourth chapter, where we delve into the origin and development of biodiversity, concentrating particularly on the ups and downs of the past 600 million years. We'll enter into a debate on the origins of biodiversity which goes to the very centre of what we think about ourselves and the other organisms with which we share this planet. Much of our attention will be on extinction, both in the past and in the present.

Up to this point in the book biodiversity is discussed, at least as much as is possible, as an objective scientific body of knowledge. But part of the reason we find it difficult to get a handle on the term biodiversity is because, in the minds of many, it is a value-laden concept. Furthermore, because we have to rely on measures of biodiversity, and the measures we pick often reflect what it is we value about biodiversity, how could the whole subject not be value laden – whether we like it or not? So the remainder of the book is devoted to the threats to, and value(s) of biodiversity, including direct and indirect monetary value. The main threats are discussed and illustrated, paying particular attention to one of the main drivers – us. We will take a path that leads us to an attempt by economists to cost the earth and its services, a project by scientists to create a living life-support system for eight people, a current scientific controversy on how many species we actually need, and a survey of religious thought on the place of biodiversity, and nature in general, in our thoughts and beliefs. The penultimate chapter leads on from talk of value, to what have we done, and what are we doing, to conserve biodiversity. The final chapter, for me personally, draws the whole book together, but in other ways it's optional. It is a personal view on what all of this biodiversity stuff means.

This beginner's guide to biodiversity is aimed mainly at those with little formal training in biology who want to find a way into some of the most interesting biology questions and some of the most pressing biodiversity issues of our time. The worldwide web is to someone interested in biodiversity what a refuse skip is to an ecologist like myself – filled with a lot of filthy or irrelevant material, often of dubious worth, but sometimes containing really neat stuff that, once salvaged under cover of night, can be extremely useful for your research. For this reason I have included some references to websites (marked in the text by a 'www' superscript) which the reader can use to 'go further' and pursue in greater detail some of the material that we can only skate over the surface of here. In my experience it seems much easier and more convenient to direct people to a website than ask them to get hold of a book or an article. But knowing where to look for good stuff is everything, and some direction as to why it's good and relevant is invaluable. I hope that I'll give the reader a few starting places. At the end of the book I'll also list just a few 'must-read books': but this is a beginner's guide, an attempt to get you interested, so the books and websites are hardly exhaustive.

I've tried to make what follows quantitative rather than go for the 'ooh-ah' factor. After all, the bread and butter of science is what you can measure or quantify in some way. But that is not to say this is all that I personally value about biodiversity – facts and figures, calculations and guestimates. I have felt the wonder of peering down a microscope for hours on end watching an embryonic shrimp or snail develop, seeing the separation of its cells, witnessing its first heart beat. I have been overwhelmed by the beauty and complexity of the living world, from tens of different little microscopic creatures inhabiting and working their sandgrain 'planet' to patterns of life spread majestically across a much bigger planet. But this is not the time, or the place. The majesty and wonder of biodiversity is always better 'felt than tell't'. Just now we live at a time in history where, I will argue, our living world will at best diminish, at worst disappear. It is good to try to understand the facts of the matter, to help inform us about how we feel about it, and what we should do about it.

For some, there will not be enough rigour here. They will want a more balanced, more detailed, more comprehensive (there's lots of stuff that many would consider essential that I don't even mention, let alone discuss), less personal account and discussion. They will want a much more academic approach. To those people, I suggest putting this book down and buying a copy of the textbook *Biodiversity. An Introduction* written by Kevin Gaston and myself.[www#4] Admittedly I'm biased, but I think it's pretty good for such a comparatively small book. Alternatively, you could read Christian Lévêque and Jean-Claude Mounolou's *Biodiversity* or Mike Jeffries' textbook *Biodiversity and Conservation* (1997). They too are good, but in different ways.

For those of you still with me, we'll start by asking how many different living things there are on the earth and how they are related.

chapter two

all creatures that on earth do dwell

If I have told you these details about Asteroid B-612 and revealed its number to you, it is on account of grown ups. Grown ups love figures. When you talk to them about a new friend, they never ask questions about essential matters. They never say to you: 'What does his voice sound like? What games does he prefer? Does he collect butterflies?' They ask you: 'How old is he? How many brothers does he have? How much does he weigh? How much money does his father earn?' It is only then they feel they know him.

The Little Prince, Antoine de Saint Exupéry

the big picture

To think of life on our planet only in terms of facts and figures, percentages and ratios is, to most of us, I imagine, simply 'not enough'. It is too narrow a view of 'how things are'. Materialists and our rock legends may tell us that all we see is all there is, but intuitively we seem to know that there is more to it than that – even if we're not entirely clear what we mean by 'more'. And yet, if we are to come to some general consensus about the 'big picture' with regards to the living creatures with which we share this planet and what we are to make of them, we must first attempt to stand back and consider the best facts – not just stories – available to us. We need to consider them in as objective a way as possible. But let's be clear. I've already said I'm not suggesting that what we should value most is the putting together of a dispassionate catalogue of living things. Far from it. But we must not allow the many things that have been, and are, *believed* about life, about biodiversity, to 'cloud' our judgement at this point – not yet, at any rate.

In this chapter and in the two that follow we set about a short scientific study of life on earth. What we want is to see clearly the 'big picture'. Ironically enough, it is actually quite difficult to get this big picture – even from very good biology textbooks and monographs. Biology courses at school and university too often focus on detail – and to such an extent that you need a thorough scientific training to follow, let alone attempt to understand, what is going on. So, is it possible to produce a 'big picture' that can be read and understood by anyone who is interested? I do hope so.

We'll start with a brief look at the earth itself and then consider in a bit more detail the living things that inhabit it, how many different types of living beings there are, or have been, where they live and what they do.

the ball in my garden

For most of us, getting to grips with things that are just too big or too small to visualize is a problem. How do we get some perspective on something as mega as 'life on earth'? Let's start in my garden. A small, children's-sized soccer ball lies almost hidden by a blackberry bush. The ball is about 25 cm in diameter. Imagine that we could shrink the earth to the size of this ball. As our planet is actually 12,756 km in diameter it's roughly a $\frac{1}{51}$ million scale model. The distance from pole to pole is about 42 km less than this, so imagine using your foot to squash the sphere slightly by about 8.2 mm and you have a good idea of what the planet looks like. Just under a metre away there is another ball about a third as big as the original ball. This represents the moon, 384,400 km from the earth. On this scale the sun is a giant ball, over 27 m in diameter, sitting in spectacular fashion somewhere in the nearby village of Hollacombe about 3 km away and on the main road to Plymouth. The planet Pluto, at the furthest reaches of our solar system, would be represented by an object, less than a quarter the diameter of the football, sitting 116 km away just south of Bristol. The nearest star, Proxima Centauri at 4.22 light years away, on this scale would be one-fifth of the way to the real moon. For the next planet orbiting a sun like our own (exoplanet HD217014 orbiting 51 Pegasi), we're talking about a fairly big ball twice as far away as our moon.

Now let's return to the garden. We pick up the ball and look at it. It's been lying there all through the winter and spring – it is, as I write, now the beginning of summer. You would see that the bright whiteness of the ball is obscured over most of its surface by a green and living crust – a green slime. A fingernail drawn ever so gently across the surface is enough to slice through the green covering – only marginally thicker than the page you are currently reading – and reveal the true colour of the ball. And yet this living shell on the ball is proportionally many, many times thicker than the space within which every living being can and does occur on earth. All of life on earth – and, as far as we know, life of any kind – can be found within a band about 25 km thick across the earth's surface: that's half of one millimetre on the football scale of things. The region where weather happens (the troposphere) is on average 12 km high, 10 km at the poles and 16 km at the equator. So between zero and 10 km high is where we find all flying and nearly all land-living life, and even then it is mostly concentrated at the bottom. I say nearly all land-living life because there are living creatures found beneath the surface of the earth, in caves or in soils; in fact, some very simple forms of life have even been recorded from 4 km deep underground. Water covers about three-quarters of the earth's surface. The average depth is 3.9 km, with the deepest part 11 km and there's life here too.

This thin skin of all the living beings on earth (referred to as the biosphere) really is quite small and fragile in the great scheme of things. Before we go on and look at how many species there are, and what they are, we first have to ask the question, what is this thing we call a species?

what is a species?

Species are the common currency of life on earth. The way we use the term is fundamental to how we understand biodiversity. It is difficult to talk about any area within biology without referring to species in one way or another. And yet even this is a problem. Jean-Baptiste de Lamarck (1744–1829), an early evolutionist (right process, wrong mechanism) in his famous study *Recherches sur l'Organisation des Corps Vivants*, wrote: 'The more our knowledge has advanced, the more our embarrassment increases when we attempt to define a species. The more natural specimens are collected, the more obvious it becomes that almost all gaps between species are filled and our dividing lines fade away.' No surprise then that there is still much debate over this most elementary of all labels. Currently there are at least seven different definitions.

When most people use the word species more often than not they are talking about the *morphological species.* Telling species apart has traditionally depended on noting differences in what similar forms look like. For example, off European coasts there are a number of fish that look like herring: the sardine, the sprat and the herring proper. They share a herring 'shape' but there are differences you can plainly see. And these differences, as long as they are consistent and if they breed true, can be used to separate the different species. For example, the lower jaw of both the herring and the sprat is consistently longer than the upper jaw, whereas the jaws of the sardine are of equal length. And the belly of the sprat has a distinct serrated or saw-like edge and is strongly keeled, compared with the herring which is weakly keeled and not strongly serrated. So biologists use a biological key to identify species. A key consists of such differences often organised into the type of question/answer approach that you, for instance, might use to diagnose problems with your washing machine. Start at first question – Is the 'On light' on? Yes – go to next question. No – try plugging it in, stupid. Question 2 – Is the drum spinning? And so on. A simple key for discriminating between our three members of the herring look-alikes could be: Question 1 – Is the lower jaw the same length as the upper jaw? Yes – then it's a sardine. No – go to Question 2. Question 2 – Is the belly serrated? Yes – it's a sprat. No – it's a herring.

Those of you who have seen a herring and a sardine may say, 'The herring's dark blue on top and the sardine's a sort of greenish-olive colour; is that not more obvious?' Well, yes it is – as long as the creatures are alive or newly caught. But as is the case with many biological collections, identification is often carried out on pickled specimens, particularly if you have a lot of stuff to identify. And the stuff used to pickle animals, formalin or alcohol, quickly removes colour and life-like appearance. Choosing features that don't change after preservation, such as relative lengths or numbers of structures, is usually seen as the best option. More recently differences in micro-structures (that is, some of the common, information-containing, molecules of species – the genetic substances DNA and RNA) have been used to successfully classify and identify species. Such differences accumulate in each of two species (derived from the same ancestor) as the amount of time they are separated increases.

Now, there are other scientists who think that even if two individuals are near identical in appearance, but live so far away from each other that there is no chance of them getting together (in the biblical way), then they should be considered as separate species. So only an interbreeding bunch of neighbours would be referred to as a species, in this case a *biological species.* For example, Audubon's Warbler, with its distinctive yellow throat, and the Myrtle Warbler, with its distinctive white throat, were originally described as two separate morphological species. However, it was subsequently discovered that their breeding ranges overlap in Alaska and north-west Canada, and here they successfully interbreed. So what were two species are now considered as one, only, the Yellow-rumped Warbler. One problem with the biological species concept is that because it is so reliant on sex, creatures that do not have sex, like little water fleas and many groups of microbial life, don't quite fit.

A third definition of a species is the *evolutionary species*, a single 'blood line' of ancestor-descendants, with its own evolutionary path and history. To take another bird example, there are a number of different kinds of red crossbill which can occur together, but quite literally mate only with their own kind. Each of these different kinds would be an evolutionary species. The biological and evolutionary species, together with the cohesion, ecological, phylogenetic and recognition species are all interesting ideas on how to define a species. But the bottom line is that, currently, the morphological species is the easiest and cheapest one to use, and so will probably continue to be the most commonly used for large-scale biodiversity studies.

For the last couple of hundred years biologists have agreed among themselves to give species two-part scientific names (a binomen). These names are always Latinized following the custom prior to the eighteenth century of publishing scientific papers in Latin. Latinized species names now provide a universal language for the naming of living creatures. So the blue whale has the scientific name *Balaena musculus.*

It may come as a surprise to the non-specialist but no one actually regulates the names given to species. Often the name chosen reflects the imagination, understanding and/or perversity of the person(s) who decides upon the name. For example, the scientific name of the narwhal (often referred to as the unicorn of the sea as it is a whale with a huge horn protruding from its head) is *Monodon monoceros.* Carolus Linnaeus (1707–78) – a Swedish biologist who was the first to suggest and use the binomen – originally described it and referred to it in his work as 'one-tooth, one horn'. The Latin name is made up from three Greek words: *monos* = single, *odontos* = a tooth, and *keras* = a horn. However, not all species have such 'serious' names. *Agra vation* is the name of a tropical beetle that was very difficult to collect; *Crepidula fornicata* is a type of limpet with very distinctive reproductive behaviour; *Massisteria marina* is a little, one-celled, seawater dweller which caused some commotion when it was discovered; and *Traskorchestianoetus spiceri* is a tiny mite living on the belly of little shrimps, and given that name despite the protestations of its discoverer.

how many living species are there ... and what are they?

Exactly how many different types of species are there alive on earth today? To be truthful no one really knows because no one has gone out and counted them all. Serious estimates vary between 3,635,000 and a staggering 111,655,000. The best and, for the majority of scientists, most realistic estimate is that there are 13.62 million species. If this is so, then only a tenth of all species have been formally described.

Often when people talk or think of living things they have in their minds people, dogs, cats, tigers, pandas, gerbils and even perhaps lizards, frogs, toads and fish. These are all animals, and relatively large animals at that. But biodiversity is so much more than just large animals, which, as we shall see, comprise only about 0.4% of the different species.

To get an idea of the different kinds of life that make up those 13.62 million species, we return to Wembury Bay. For in the life we find along this small area of beach we have, to some extent, a picture of what we find globally (Figure 2). Introducing the different groups we find on this one beach gives us a way into discussing life on a global scale.[www#5] Below is a 'top ten' of the most diverse groups of living things worldwide – diverse, that is, if we take the number of species as our measure of diversity. The groupings are natural but do not fit neatly with any particular level within a more scientific classification (see Appendix). We'll also take a little time to consider why these species are seen as (economically) important to us, remembering that we will discuss how we do and/or should value biodiversity in a later chapter.

no. 1 nearly every species has jointed legs

To the nearest approximation every species on the beach at Wembury, and worldwide, is a jointed-legged animal with a hard outer skeleton, an arthropod (from Greek *arthron* meaning 'joint' and *podus* meaning 'foot'). Below the high-tide mark the

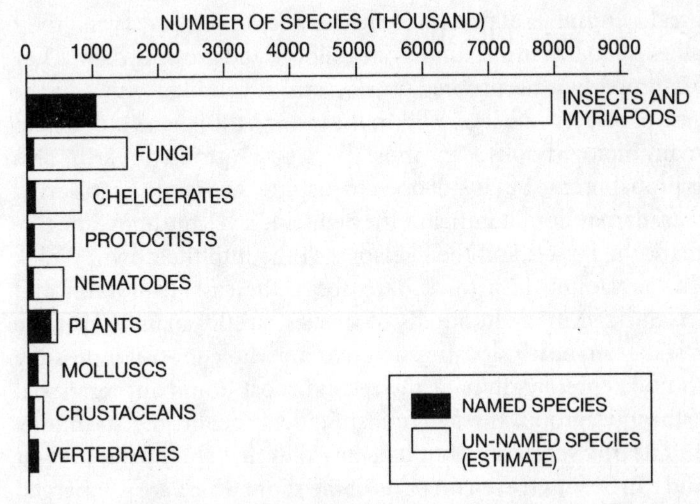

Figure 2 *Number of species on earth (information from* Millennium Ecosystem Assessment, *2005)*

crustaceans, the so-called 'insects of the sea', dominate. Turn over any rock and there are shore crabs or, lower down, aptly named 'angry crabs' with deep blue, velvet-covered shells and bright-red eyes. Look in any pool and you will see, if you look carefully, the small glass prawn or its big brother the edible prawn. The white band that looks from a distance as if someone has painted a strip horizontally along the shore, is made up from the outer shells of many thousands of tiny barnacles. In the pools, beneath the rocks and in the gravel and sand are hundreds of different types of microcrustaceans, mainly copepods – but you would need a good hand lens, or even better a microscope, to see these. There are an estimated 150,000 crustacean species worldwide, with only 40,000 currently described. However, as we shall see, this figure, though impressive, pales in comparison with their more land-based relatives.

By far the greatest numbers of species on earth belong to the land-living arthropods, the insects, centipedes and millipedes,[www#6] eight million of them, or 59% of all species, although there has been a lot of debate recently as to whether we're considerably overestimating our estimate. Certainly the accolade of the greatest number of described species goes to the insects and their allies – 950,000 in all, almost one million, and growing daily. They are properly waterproofed crustaceans, if you like, with – in the case of insects – six legs. Within the insects, the beetles are cited in many biology books as the most species-rich group on earth. This is probably true but is still open to dispute. At Wembury the crustaceans may dominate below the high-tide level, but above the tide mark the insects and their relatives come into their own. Wrack flies are found in a thick haze above the cast-up flotsam and jetsam, and many hundreds of species, beetles mainly, visit the wrack from nearby locations. Across the whole shore it is possible to find representatives of the second most dominant terrestrial arthropod group, the arachnids. Spiders inhabit the strandline and nearby vegetation, but it is the mites that are superabundant and can be found even on parts of the shore which are submerged underwater for considerable periods of time. Worldwide there are an estimated 750,000 species of arachnid, but with only one in ten of these actually described.

The arthropods are of huge economic importance to us. The great majority of insects are 'plant-eating' (herbivorous). They eat more plants than any other creature on the planet. So they include major pests of nearly every one of our major food crops (e.g. the Colorado potato beetle, which attacks potatoes, tomatoes and egg-plant, and the Yellow stem borer, which is the most pervasive insect pest of rice), as well as species that attack stored foods and the wood we use to build furniture, houses and the like. The costs of damage caused by such pests can be astronomical. Take the current problem with subterranean termites in the state of Georgia, USA, where damage to homes is estimated at US$17 million, with a control cost of US$58,600,000. This said, less than 1% of insects are pests, with only about 100 species being persistent problems. Other species feed on such pests, such as ladybirds feeding on aphids. And many crop species would not be able to reproduce without insects to help them by carrying pollen from one plant to another. Those species that live in the soil, together with those that scavenge, are critical in breaking down and recycling organic waste. There are some non-herbivorous insects, some of which (at best) are irritants to us and our domesticated animals, by biting and sucking – those readers who have spent a calm summer evening beside a Scottish loch will be well acquainted with the 'midge', an extremely tiny, irritating, biting fly. At worst, these biters and suckers spread disease; bubonic plague spread by fleas, malaria and yellow fever by mosquitoes, are just a few convincing examples. Some species of ant, beetle, caterpillar and locust are actually eaten by people, as are some insect products, such as honey from bees.

The arachnids contain many poisonous species of spiders and scorpions (some of whose bites can be fatal), but perhaps the most important economically are the ticks and mites. Like some insects, they can parasitize, infect or just irritate people and their domestic animals. Many crustaceans such as crabs, lobsters and shrimps are an important food source, but we should not forget that it is many of the tiny microcrustaceans that form a vital component of freshwater and marine food webs.

no. 2 mushrooms and moulds

After the arthropods, the Fungi are the next most species-rich group.[www#7] These are mushrooms, moulds and yeasts. On Wembury beach proper there are very few fungal species. They are much more land-living than marine, although you will find some species skirting the green rim of the shore, as well as on, and in, some of the marine animals present. We know remarkably little about land Fungi, and even less about their poorer relatives, the marine Fungi. Just travel a few miles east from Wembury to Slapton Ley and there, in an equivalent-sized coastal area, around 2000 species of Fungi have been recorded. Of an estimated 1.5 million species about 72,000 have been described. In contrast with the case of insects, some researchers believe this 1.5 million to be an underestimate and place the figure as high as 2.7 million. However, the bottom line is that the group is comparatively poorly known, making predictions very difficult. Many people make the mistake of lumping Fungi with plants. Ironically, they seem more closely related to animals than plants, and are so different from both that they are rightly placed on their own, as we shall see later.

Many Fungi cause diseases, but many more form constructive, intimate associations with plants. Such Fungi are near essential for plants to grow in nutrient-poor soil. Moulds and yeasts are used extensively in the production of cheese, beer, wine and soy sauce. Antibiotics, such as the mould Penicillin, are now a key area of health care. Penicillin mould also develops flavour in Stilton and other blue cheeses. Arguably, the largest creature alive on earth today is a fungus called *Armillaria.* There is a living clonal growth in Washington, USA, which covers an area of 600 hectares and one in Michigan, USA, which weighs more than 100 tonnes. That's about the same mass as a blue whale. Both of these Fungi are more than a thousand years old. Of course, most Fungi are much smaller, and often we are not aware of them until we see their fruiting bodies, what we call mushrooms or toadstools, or until they turn the timber in our house to mush through dry rot.

no. 3 mini-life

Only around 4000 species of bacteria have formally been described.[www#8] Most of these belong to one group, the cyanobacteria. So why is this grouping third in our list of most species-rich groups? Well, when I asked a microbial biologist colleague of mine how many species of bacteria he thought there were on Wembury beach (he has carried out some of his research there) his reply was, 'Depends what you mean by species ... different kinds, about 10,000 ... you're not going to quote me on this are you?' So it should come as no surprise that in terms of estimated species number the world total for bacteria is around one million. Still, as the species guestimate for Wembury suggests, even this global estimate could be way out, with figures of three million or more currently being banded about. Certainly, one major problem with such mini-life is how to recognize a species. We said earlier that morphological species were by far the easiest to use in most circumstances. But this is certainly not true for such tiny critters where, in one sense, there is very little to see. Bacteria are very 'basic' to look at, often being sphere or rod or spiral shaped. So biologists have resorted to chemical tests to supplement the differences in what particular bacteria look like. Even then, you need to be able to grow the bacteria in the laboratory to test it and most bacteria simply cannot be grown under laboratory conditions. More recently, observation of differences in the composition of molecules such as DNA has been used to discriminate between different species, and this technique is proving extremely powerful and useful.

Bacteria are nearly all very small, microscopic creatures between 1 and 5 millionths of a metre in length. The largest species is 0.75 millimetres in diameter. Discovered in 1997, it is a marine bacterium *Thiomargarita namibiensis*, the sulphur pearl of Namibia. It gets its name from the fact that it uses sulphur in its life processes and inhabits coastal mud off Namibia, Africa. About half of the species known are capable of moving in a particular direction, and at speeds up to x100 their body length per second. They are extremely numerous. A spoonful of garden soil contains more than one thousand individuals. I once read that the total number of bacteria in a person's mouth is greater than the number of people who have ever lived. One tenth of our bodies' dry mass is made up of bacteria. It has been estimated that the world contains 5,000,000,000,000,000,000,000,000,000,000 individuals and this number is growing fast. Bacteria can be free-living, inhabiting the surface of other living things, as well as making their home within animals, plants and other bacteria. They seem to be important to us for many different reasons: they are essential to the health of digestive systems, in soil maintenance in agriculture and forestry, and for the existence of the air we breathe. Truly, the bacteria are the hardiest of living beings. Some can survive extreme temperatures including boiling-hot springs.

When thinking about bacteria we come to one of the most fundamental divisions we find in living things. Together with a much less species-rich group of very primitive microbes, the Archaea, the bacteria constitute one of two fundamental units of life, the prokaryotes. Everything else that is alive – the animals, the plants, the Fungi, the one-celled microscopic creatures – all belong to the second main division, the eukaryotes. Prokaryote cells are smaller (0.2 to 10 millionths of a metre) than eukaryote cells (10 to 100 millionths of a metre), and do not have smaller organized structures inside them. Prokaryotes do not have mitochondria, which are the powerhouse of the eukaryote cell, or plastids, which are structures that plant-like creatures have where photosynthesis (fixing energy from sunlight) takes place. There are many other differences but perhaps two of the main ones are: (1) in prokaryotes the genetic material which carries the plan for building new individuals is a single strand of DNA, not contained in a cell nucleus, whereas in eukaryotes the genetic material is a combination of DNA, RNA and protein arranged in 2–600 chromosomes, all of which are enclosed in a membrane to make the structure we call a nucleus (the word eukaryote means 'true nucleus'); (2) in terms of how they get their food and make energy available to do things (metabolism) the prokaryotes, and so the bacteria, are probably the group that shows the greatest diversity. We will come back to this fundamental division between prokaryotes and eukaryotes when we consider how the members of our 'top ten', and everything else, are related to each another.

no. 4 strange bedfellows – amoeba and seaweeds

What do a microscopic slime mould and a giant seaweed have in common? One answer, neither of them can drive a tractor, sounds infinitely more believable than the true answer – quite a lot really. Incredibly, they are remarkably similar in the ways in which they are made, the ways in which they work, and in their family ties.

On the beach at Wembury there are well over 100 different species of large algae, the seaweeds. At the very top of the rocky part of the shore there are distinctly light-green patches mainly, but not exclusively, marking the flow of freshwater run-off from the land onto the shore. Below them are bands of brown seaweeds, among them saw-toothed wrack, often so dense that they entirely cover the rock beneath from view, and at the very lowest part of the tide are swaths of the majestic oar-weeds. By far the most species-rich group are the red seaweeds, finely sculptured, and often incredibly intricate and stunning forms, found low down on the shore, predominantly in rock pools and in protected crevices and overhangs. But unseen, and some orders of magnitude smaller, are tiny algae and what used to be referred to as one-celled creatures, the protozoans, living on and between the rocks as well as in and around sand grains on the sandy beach. The number of protozoan species on Wembury beach has not been counted for about fifty years. Then there were fewer than 100 species present.

The new grouping, which encompasses the protozoans and the algae as well as slime moulds and a whole bunch of less well-known free-living and parasitic organisms, is the Protoctista.[www#9] While only 40,000 species have been described, the estimated species number for the protozoans is 200,000 and for the algae 400,000. It is generally recognized that these are pretty poor estimates, but even then they make the Protoctista the fourth most species-rich grouping. They are all referred to as 'lower' creatures, with some having animal-like and others having plant like characteristics. There are multi-celled creatures such as the seaweeds, but there are also the one-celled organisms such as amoeba. Strange characters like the slime moulds can be either one-celled or multi-celled. For example, slime moulds may be found as individual amoeba like creatures living in soil-eating bacteria. However, when the food source 'dries up', all the cells get together and form a sluglike creature (with the wonderful name of 'grex') which then crawls to a new environment. Once there, the slug is transformed into a towerlike, umbrella-shaped structure from which spores are released to colonize the new habitat.

Despite sharing family likeness, in terms of how they're constructed and how they function the Protoctista can be considered the most diverse of all eukaryote forms of life. They are found everywhere there is water. In salt water and freshwater, living on the bottom or as plankton (free-floating inhabitants of the sea or lakes), in damp soils and leaf litter, and inside every other living thing, including you and I.

no. 5 roundworms

Only eight species of nematode roundworms have been recorded from Wembury beach, nearly all from the holdfasts, the attachment structures, of seaweeds. The main reason for such a low figure is twofold. Nematodes are notoriously difficult to identify and those who know how to identify them have never carried out a survey at Wembury. One individual who knew how to identify them made an extensive collection at one location west of Wembury and closer to the city of Plymouth, Tinside. He found almost seventy species in a very small area.

So, given the difficulties of a group where peering at their genitalia through a high-powered microscope is an essential part of identification, it should come as little surprise that although only 25,000 species have been described, the currently accepted estimate is 400,000, with final figures varying between one million and one hundred million appearing in the scientific literature. Certainly the accepted estimate is poor, but the idea that a section of deep sea floor has 100 million species, and so is more diverse (in terms of numbers of species) than a rainforest, is now waning a little.

Nematode roundworms (or thread worms or hair worms) look on the surface a bit like earthworms but with no segments.[www#10] Inside they show none of the repetition of structures that we see in earthworm segments. When very small they look like individual threads, hence their name (from Greek *nema* meaning 'thread'). They have a tough outer skeleton and an inner muscle arrangement, which means they thrash about a lot. They are extremely prolific, with some species able to produce 100,000 fertilized eggs every day. Someone has estimated that there are 100,000,000,000,000,000,000 individual roundworms on earth, and this doesn't take into account species that are parasites of plants and other animals. Most free-living species are quite small but this is not so for all the parasitic species, where centimetre, and occasionally metre, is the measurement unit. The largest known roundworm was found in the placenta of a female sperm whale. It was 9 m long.

Probably the best-known roundworm is the soil-living *Caenorhabitis elegans*. As part of the Human Genome Project, the genome of this 'representative invertebrate' (animal without a backbone) has been worked out and published. With a size of 80 Mb, this makes it one of the smallest animal genomes worked out to date. When it comes to roundworms known for bad reasons, the possibilities are endless. A reasonable number of different species parasitize and damage key crops such as cereals, cotton, citrus and other fruits, legumes, potatoes, sugar beet, trees and so on. While there is a similarly large list for animal parasites, it is in humans and their domesticated animals that the effects are, sometimes literally, felt. There are a number of tropical, blood-sucking intestinal hookworms: for example, *Ascaris*, which lives in the small intestine where it 'interferes' with the host's ability to take up food, and *Wuchereria*, which blocks the lymph channels resulting in the 'disease' elephantitis. Many parasites require a carrier animal to infect the person, or the pig, or the horse. For instance, the roundworm *Onchocerca* is transmitted to humans by blackflies and infection results in tropical river disease, and in some cases 'river-blindness'.

no. 6 viruses

Up to this point all of the life forms described are either cells or are made up of a number of cells and are 'living' in the sense that they have all of the machinery required to replicate and carry on the family line. One form that does not quite fit the 'living' category is the virus.[www#11] Each virus is composed of quite literally bits of genetic code (like DNA) wrapped up in a protein coat. They seem to be more closely related to the cells in which they reproduce than to each other. And it is the fact that they return to already living tissue, and are dependent upon access to its complex structures and chemicals to enable them to replicate, that marks viruses off as something quite different.

When it comes to what viruses there are on Wembury beach it is difficult to know where to start. Certainly, in the last ten years the discovery of a huge diversity of marine viruses has transformed our understanding of how the oceans work. So there are bound to be many species of virus in the seawater in the bay. Indeed, as has just been discovered, it is the viruses that give seawater its distinctive 'sea smell'. And every living thing on Wembury beach is most likely to have its own complement of specific viruses. A couple of weeks before I wrote this section one of my students detected the presence of a couple of viruses on weed cast up in the strandline, and also in and on some beach-hoppers that were his primary interest. If you look, you find. But at present we have no real idea of how many viruses are present on Wembury beach. The situation is not much clearer for our global inventory. Four thousand viruses have been described, but the estimated species number is a hundred times that. And even that is a very conservative estimate. The fact that, as I said above, viruses seem to be more closely related to the species they are found in than to one another means that as a minimum estimate there could be as many viruses as there are species – thirteen million odd. This assumes, of course, that each species has 'produced' only one virus. Given how little we know about species diversity of viruses, we are quite knowledgeable about their genetic diversity. The genome (all of the DNA taken together) of many hundreds of viruses has been mapped. This is principally due to the small size of their genomes. Hepatitis B is one of the smallest genomes at 6.4 thousand nucleotide bases (Kb) and Smallpox one of the largest at 372.2 Kb. Compare these with the genome for the mouse (6,013,260 Kb) or even the humble fruitfly (137,000 Kb).

no. 7 greenery

For our seventh most species-rich group we have organisms that, unlike much of what has gone before, should be instantly recognizable, the plants or the Plantae.[www#12] These are the mosses, the ferns and other spore- and seed-bearing plants. There are examples of all of these skirting Wembury Bay, well above the highest tide level. On Wembury beach itself there are very few plants; remember, seaweeds are not plants. Just around the corner in the Yealm estuary, below the low water mark, is a bright green 'meadow' of one of the only marine flowering plants, *Zostera*, the sea grass or eel grass. But there is no *Zostera* in Wembury Bay. Around the high water mark, though, it is possible to find salt-tolerant flowering plants such as Sea Rocket or the Yellow-horned Poppy. However, we are in the wrong environment if we really want to study plant diversity. Most plants are land-living and are found in deserts, forests and grasslands, with very few returning to the water.

Up until now the gulf between the number of species we have described and how many we estimate are 'out there' has been colossal, with our estimates ranging from reasonably poor to very poor. However, we do have a pretty good idea of how many species of plant there are. There are currently 270,000 described species, with a fairly good estimate of 320,000 species in total. Part of this is owing to the fact that plants, for the most part, are large and stay put, and there has been a huge and long-standing interest in them as sources of food, building materials and medicines (see chapter 5). To the nearest approximation, every plant is a flowering seed plant, an angiosperm. The next two largest groupings are the seedless ferns and horsetails and the spore-producing mosses.

no. 8 living homes exhibition

Walking onto the foreshore at Wembury there are a number of things which immediately strike you. The amount of green and brown seaweeds, particularly in the summer, define the shore. But so too does the sheer number of shelled animals on the surface of the rocks. There are many hundreds of dogwhelks, many thousands of limpets and many tens of thousands of periwinkles and topshells. And on this beach the shelled animals are the second-most species-rich animal group. They are so obvious. These shelled animals belong to the Mollusca (from the Latin *molluscus*, 'soft', referring to their bodies not their shells), a group comprising snails, clams and cuttlefish.[www#13] In the summer, snorkelling in the shallows of Wembury Bay you can see many cuttlefish, looking for something to eat. All molluscs have a muscular foot attached to a soft body containing the internal organs. There is a fold of tissue (the mantle) draped over the main body mass, which is responsible for producing a hard shell made from chalk (calcium carbonate). Most species of mollusc are marine, but they are also very successful both in freshwater and on land. The estimate of 200,000 living species is considered quite reliable, although to date only 70,000 species have been formally described. Large specimens, primarily bivalves (clams and the like) and cephalopods (squid, octopus), can be measured in metres but some of the smallest require a microscope to see them, and there is everything in between. Squid, oysters, scallops, cockles, periwinkles and land snails are highly prized as food, while some species and their products are considered as natural objects of beauty, e.g. mother-of-pearl and pearls themselves. It's not all positive, though, as some species are pests of farmed bivalves. Also, the shipworm, *Teredo*, effectively destroys untreated submerged wooden structures or boats and many snails act as intermediates for the infection of other animals, including humans, by parasites.

no. 9 animals with backbones

The most common and species-rich backboned animal (vertebrates) worldwide and at Wembury are, arguably, the fish. There are many species which specialize in life between the tides, like the blenny and the Cornish lumpsucker, but there is also a host of other species which use the warmth of the upper rock pools as a 'nursery area' for their fast-growing young. However, the bird life too is quite spectacular. Birds of prey swoop down from the National Trust car park above the beach, which is littered with various gull species and others such as wagtails and crows.

Living vertebrates are made up from about seven different, though related, family lines.[www#14] Three of them are completely water-dwellers and have evolved a fish-shaped body form independently from one another: the jawless fish, such as lampreys; the cartilaginous fish, the sharks and the rays; and the bony or ray-finned fish, the pipefish, stickleback, trout, wrasse, perch, flounder and the like. The most species-rich group belongs to the perch family. To the nearest approximation, every bony fish is a perch and so too is every fish. In fact, about half of all vertebrate species are fish, and bony fish at that. There is an estimated 50,000 vertebrate species, with 45,000 already described. It is believed that, as with the plants, this estimate is quite accurate.

There are three groups which are, for the most part, not aquatic although they have some members that have returned to the sea; the reptiles, the birds and the mammals. For the group we belong to, the mammals, most of the groupings it contains are simply not very diverse, with few exceptions. To the nearest approximation, every mammal is a rodent. Straddling a watery and a land existence are the amphibians. Their aquatic young stages have a typical fish form, where the land-living adults more closely resemble the reptiles and mammals in their body form, although, unlike these two groups, adult amphibians are poorly waterproofed.

no. 10 everything else

All of the remaining groups, encompassing well over a hundred different body designs, lumped together comprise an estimated quarter of a million species. Currently only 115,000 have been described. However, just because these species are awarded joint tenth place may not necessarily tell you anything apart from the fact that most of these groups are not very species rich. And taking a different measure of biodiversity may well have yielded a different ranking order. Certainly, some of the most noticeable, and noteworthy, animals on Wembury beach have not made it into the top nine. For example, the underside of many of the rocks are carpeted with the white outer shells of many hundreds of filter-feeding worms and sea mats living in colonies of hundreds of individuals. There are also sea squirts, relatives of the vertebrates, encompassing solitary forms like the lightbulb sea squirt *Clavelina* (it really does look like a light bulb), or colonial forms such as the beautiful blue star sea squirt. Sponges, the most primitive of all animal forms, but with internal support provided by a bewildering diversity of often incredibly beautiful and intricate little structures called spicules, star-shaped, pen-shaped, even anchor-shaped, can be found everywhere. And how could we miss the little cushion star, the species *Asterina gibbosa*, belonging to a group called the spiny-skinned animals, the echinoderms. The grouping has only 6000 species but shows incredible diversity, from the flower form of the ancient sea lily, through to the sea urchin and starfish, ending up with the peculiar, elongated, sausagelike sea cucumber (H. G. Wells referred to sea cucumbers as 'nasty green warty things, like pickled gherkins, lying on the beach' in *Aepyornis Island*, 1927). In fact, if we were looking at the biodiversity of the deep sea, in terms of animal life the echinoderms would be up there in our top three. In conclusion, the measure of biodiversity we used in the first section of this chapter (species richness) precludes us from talking here about diversity at higher levels of biological organization, i.e. above the species level. That will come later.

new species

We are still a long way off having a complete catalogue of all the living beings with which we share this planet. Species continue to be described but not in any systematic way. The groups that receive attention do so as a result of individual scientists and amateur enthusiasts choosing to work on their favourite groups. So most new descriptions are of insects and spiders, groups which command a wide interest from professional and amateur alike. The main exception to this is when a group of organisms are of medical or economic interest. By and large, the majority of new species described are relatively common (not rare), comparatively large (not microscopic) life forms that are collected mainly from temperate (not tropical) regions. About one-and-a-half new species are described every hour (13,500 each year). But not all new species are large, common and/or belonging to obscure groups. New fish species are being described at a rate of just over ten a month. With birds it's one species every two months. Even within the mammals, about one new species is described every three years, with a relatively high proportion being aquatic forms such as whales and dolphins.

Description of new species is not uncommon even in areas where the wildlife is thought to be well known. But still, the rate of description is comparatively low. Indeed we may be facing a situation where some groups of species are becoming extinct faster than they can be described. Putting together catalogues of species is not hampered just by the rate of species description but also by the fact that sometimes different species are given the same names, and sometimes the same species is given different names in different locations. Although these two activities are bound to cause problems we still do not know the extent to which each is a significant problem. It has been shown that one-fifth of insect species have more than one scientific name. And it's not just at the level of species description that we see change and reconsideration. As we shall see next, the higher classification of living things is still far from decided and is still the subject of much debate.

a place for everything and everything in its place

The process of classifying species (describing them and putting them into a larger framework) and the product of this process are referred to as the study of systematics or, to use an older word, taxonomy. Thus, taxonomy is one of the mainstays of biodiversity research.

There have been numerous attempts to place all the elements of the living world into meaningful, discrete and useful categories. The philosopher/poet/biologist Aristotle (384–322 BC) came up with what by the middle ages had developed into 'The great chain of being'. We talk today of holistic and inclusive approaches to a subject but Aristotle's scheme, and its derivatives, reminds us that this is far from a modern craze. His classification scheme included what he believed to be all life, natural and supernatural. In his *Study of Animals* he says: 'Nature proceeds little by little from things lifeless to animal life ... next after lifeless things in the upward scale comes the plant ... there is observed in plants a continuous scale of ascent towards the animal ... [a]nd so through the entire animal scale there is a graduated differentiation in amount of vitality and in capacity for motion.' One of the most interesting points about 'The great chain' is that it is not just a set of helpful box labels that can be easily retrieved when required. It is, in fact, a 'world view' carefully constructed to show the interrelations and dependence of some elements on others.

The next major scheme we need to consider is that of Carolus Linnaeus. Dubbed the father of modern classification, he came up with a hierarchical classification of life that to a large extent still has considerable influence on how we group living things today. Linnaeus, reflecting a world view common at his time, believed that all species or kinds had been created separately by God. Thus there were no relationships to reflect. So, to a large extent, how one grouped living things was a matter of personal preference. It is a bit like working in a DIY store and being given numerous types of nuts, bolts, washers and so on, to put on display. It is a good idea to arrange them in some sort of order, according to thread size, geometry or function, but there is no one way to do it. In the same way, organisms were placed into artificial categories based purely on how they looked. It was Linnaeus who introduced the terms kingdom, class, order, family, genus, species. Thus the species of housefly *Musca domesticus*, belongs to the genus *Musca* in the family Muscidae. The Muscidae is one of many families that belong to the order Diptera, which contains many different groups of fly. The Diptera belong to the class Insecta, which, as we've seen, is a large grouping that holds the beetles and butterflies among others. The Insecta sit within the phylum Arthropoda, which includes all insects, spiders, mites and crustaceans, and the Arthropoda belongs to the kingdom Animalia – all of the animals. So the little kelp fly, *Coelopa frigida*, so abundant in the weed cast up at the high water mark on Wembury beach would be classified as follows:

Kingdom:	Animalia
Phylum:	Arthropoda
Subphylum:	Uniramia
Class:	Insecta
Order:	Diptera
Family:	Coelopidae
Genus:	*Coelopa*
Species:	*frigida.*

The publication of Darwin's *Origin of Species* in 1859 heralded a new approach to classification. It was a bit like finding out that the nuts and bolts in our DIY shop were in some way related to one another, and thus how they were arranged could actually reflect those relationships. Biological classification was no longer just a case of assigning living forms a label, using a pragmatic system where easy access of material was the primary concern. Species were related to one another and they had common ancestors. So the emphasis shifted in the century that followed to elucidating evolutionary trees. These were diagrams, sometimes literally drawn by authors as trees, where species were placed at different parts on branches in a way that illustrated exactly how they were related to each other. The kingdom, phylum, etc. groupings of Linnaeus were retained but now they had to, as much as possible, reflect evolutionary affinities. Right up until the present day there still exists a tension between those who are interested in classificatory schemes which reflect evolutionary relationships (cladistics) and those who believe they 'merely' make it easy to determine the identity of an unknown organism. Arguably, one is not intrinsically any better than the other. It all depends on what you want the scheme for. If you want to put a name to a face then a carefully worked-out evolutionary scheme is perhaps excessive and may actually make what should be a simple identification process very complicated.

In comparatively recent times, the 1960s, Robert Whittaker put forward what we refer to as the Five Kingdom approach (Figure 3a). He assigned all known species to one of five kingdoms, based on evolutionary relationships determined by how they looked, how they worked and their pattern of development. Essentially, he added an extra kingdom, the Fungi, to Aristotle's scheme stripped of its supernatural elements. There were the bacteria (Monera), the plants (Plantae), the Protoctista (uni-celled organisms), the animals (Animalia) and the Fungi. We encountered all these kingdoms, or components of them, earlier in the chapter.

Even from its inception the Five Kingdom approach was the subject of much debate. To what extent did kingdoms actually

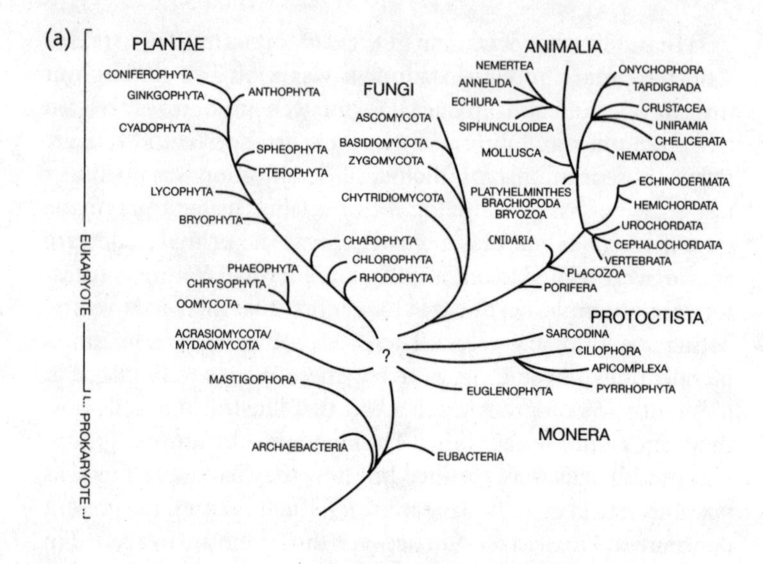

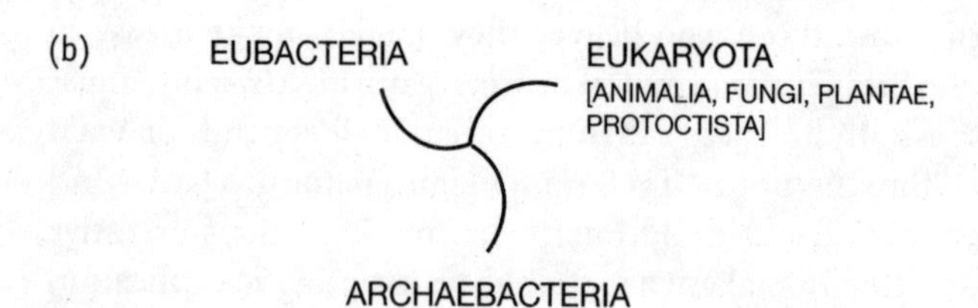

Figure 3 *The Five Kingdom model (a) and the Three Domain model (b).*

reflect evolutionary relationships and to what extent were they artificial constructs? Even the number of kingdoms that should be accepted was a matter of controversy. For example, I have a recent textbook on my shelves which advocates that the bacteria be split into two groupings – the true bacteria (Eubacteria), which would correspond with what we called bacteria earlier in the chapter, and the Archaebacteria, a much more ancient grouping – thus, in this scheme there are six kingdoms instead of five. Another textbook has superkingdoms and subkingdoms, stating that Prokaryotes are one superkingdom and Eukaryotes are another. The overall message is that there is no agreed scheme for such higher classification.

What was clear was that both Aristotle's and Whittaker's schemes emphasized 'big' (principally animals and plants) over 'small', often microscopic living things – understandable for Aristotle almost 2000 years before the first microscopes were invented. This 'size-ism' inherent in the Five Kingdom approach was challenged in the later 1970s by a scientist called Carl Woese. As a molecular biologist he tended not to associate the importance or 'worth' of living material with its size. He devised a classificatory scheme based on the evolutionary relationships of common molecules (principally DNA and RNA, therefore a classification based on gene sequencing) to one another. What he came up with really shook biologists. The Three Domain model he proposed says (in common with the six Kingdom model above) that the bacteria are in fact two very distinct groupings, the Eubacteria and the Archaebacteria (Figure 3b). Both are prokaryotic forms of life (remember, single DNA strand, no nucleus and so forth), but when the gene sequences from Archaebacteria were sequenced about 30% of the genes had never been seen before and some of them resemble human genes and those of other eukaryotes more than they do those of the Eubacteria. So there are two very early branching incidents near the origin of life, which resulted in the formation of three domains, the Archaebacteria, the Eubacteria and everything else (the eukaryotic organisms – the four remaining kingdoms, if you like).

Currently the Five Kingdoms versus Three Domains debate continues, with the subplot that even the number of kingdoms (and superkingdoms) and domains we should be talking about is itself a matter of disagreement. That the molecular evidence points to three branches emerging early in the history of life is relatively clear. What is not clear is the best way to represent this in a scheme that encompasses all of current life. In practice, animal biologists tend still to refer to kingdoms, whereas molecular biologists and microbiologists are much more in favour of talking about domains.

designs on life

The level below kingdom or domain is termed the phylum (plural phyla). We can think of phyla as a set of distinct and easily recognizable body designs. The early German biologists who contributed so much to our studies of evolution and development referred to the design that characterized a phylum as the *Bauplan* a German word which could be translated 'blueprint' or 'body plan'. We could compare different phyla with different car makers, such as Ford, Peugeot, Nissan, etc. Each company is in the business of making cars. Cars would be at the level of kingdom or domain. Each company has a suite of individual car designs (species) which fits under the broad umbrella of 'It's a Ford' (one phylum, characterized by the Ford *Bauplan*) or 'It's a Nissan' (another phylum, characterized by the Nissan *Bauplan*). After the species, the phylum is probably the next 'natural' and comparable grouping we use for classifying living things. (See Appendix for all of the organisms mentioned in this book placed into their appropriate phyla.) Comparing a species of cat and a species of Fungi, or the phylum Arthropoda with the phylum Mollusca is much more meaningful than trying to compare two genera, two families or two orders. For with the best will in the world, these levels of classification are currently very much human constructs, more so than phylum or species.

There are about ninety to one hundred present-day phyla. Some, like phylum Cyanobacteria in the Bacteria, and the Arthropoda in the animal kingdom, are very species rich. (Note that the most current view is that the Arthropoda are bigger than a phylum and contain such groupings as the phylum Crustacea – so the debate goes on.) But most are not. Indeed, on the shore at Wembury it is possible to find a large proportion of the present-day animal phyla, but for many of them we are talking about one or two species at most (maybe tens to hundreds of species world-wide). For instance, the echiuran 'worm' *Thalassema neptuni* is the only species of the phylum Echiura found at Wembury, living in holes in the rocks. There are only two species from this phylum present in the whole Plymouth area and only 150 species worldwide. The *Bauplan* of echiurans makes them quite distinct from any other creatures even although superficially they look just like a worm. We could say the same for another phylum, the Siphunculoidea: there is one species at Wembury, and 450 worldwide. It is similar for the Mesozoa, where one species commonly parasitizes the cuttlefish that come into Wembury Bay and there are only a handful of species recorded worldwide.

Why there should be so many different basic designs with so few successful at least in terms of numbers of species is still a mystery. In chapter 4 we'll spend a bit more time discussing the origins of the major phyla and how they have fared through geological time. Just now it is worth saying that within the past ten years, gene sequencing has thrown much new light on the evolutionary relationships between the phyla. What a century ago was a largely sterile and irresolvable debate about the origins of, and relationships between, the phyla, is today a puzzle that has been solved, although not every piece is yet in place.

In summary, no matter which level we look at, the kingdom (or domain) or the phylum, or even below that, life is not equally distributed between the different groupings that have been categorized. Although there are around one hundred phyla, or body plans, most biodiversity is contributed by a handful of these phyla. Most phyla are simply not very diverse. Most animals are insects, and most insects are beetles. More than three-quarters of all plant species are of the flowering variety. Most mammals are rodents. Taking all of life together, to the nearest approximation, every organism on earth is an insect.

chapter three

where on earth is biodiversity?

'What planet have I fallen on?' asked the little prince.
'On the planet earth, in Africa,' replied the snake.
'Oh! ... Then there are no people on earth?'
'This is the desert. There are no people in the desert. The earth is big,' said the snake.

The Little Prince, Antoine de Saint-Exupéry

the university of plymouth marine field course

There are a number of excellent reasons for taking a group of marine biology undergraduates from the University of Plymouth[www#15] to the Algarve, Portugal, quite apart from some excellent food and a little welcome sunshine. The marine animals and plants our students get to see and work with are often quite different from those we get on British shores. In particular, the students can observe and study fiddler crabs (remember biodiversity 'better felt than tell't'?), nervous crustaceans which run across and make deep burrows on mud flats. Apart from the birds they are the dominant 'animal feature' on the salt marsh. Fiddler crabs are predominantly tropical or subtropical creatures but the range of this particular species extends from the equator all the way north to the southern coast of Portugal. There are numerous, other, perhaps less charismatic creatures, such as the limpet with a lung, *Siphonaria*, but the bottom line is that we are introducing the students to an ecosystem made up of different characters from those back home. This nicely makes the point that is central to this present chapter. The types, and even numbers, of organisms present on earth are distributed unevenly across the earth's surface. However, there are distinct and recognizable patterns. We now briefly explore what we know of how biodiversity is distributed across the earth's oceans and land masses, identifying hotspots, coldspots and gradients as we go from the poles to the equator, from the surface of the sea to its deepest depths and from low-lying areas to the tops of the highest mountains. But first we start with a look at one of the strongest and most pervasive patterns there is, by asking the question: 'How are the number of species in a given area related to the size of that area?' And we start from where just about anyone could start – the rocky shore at Wembury.

more is more

Let's go to an area on the rocky shore at Wembury and see what we find there. In a very small area, say one square metre, on the open rock surface we find a number of barnacles all of one species, a handful of limpets, again all of one species, tiny white worms in tubes, again one species, and patches of encrusting algae *Lithothamnion*, specially strengthened by incorporating the equivalent of concrete into its body. Four species in all. If we then extend our search to three square metres, we have more work to do but the number of additional species we find makes up for the back-breaking work. On the open rock we now have two limpet species, two barnacle species and a further two species of white worm. In addition, we have found three topshell species, one periwinkle species and a solitary dogwhelk. There is also some brown algae, *Fucus*, together with some green seaweed. In a small crevice there is a beadlet sea anemone, and some sponge. We're now up to fourteen species.

We then go for an excellent coffee at the Old Mill Cafe (thoroughly recommended), right next to the beach, and wait for the second-year marine biology class to arrive. With those extra pairs of hands and eyes we can now extend our search across an area of six square metres. As well as even more bare rock, our sampling area now takes in a number of fairly deep tide pools and a couple of areas where the underside of large boulders may act as a refuge for species that we've not seen up until now. There is also a freshwater stream crossing the shore so there is the possibility of some brackish and freshwater species turning up. In two hours' searching we have totalled 208 different species. A couple of those species are rare. So what if we had extended our search to an area of sixty square metres? Most of the students would have gone home, unless the exercise was being assessed. But more important, our extended area of study would have included some of the sandy beach that makes the location such a tourist attraction, as well as the land-living animals and plants that live on the cliffs and fields that skirt the shore. The different types of environment harbour an even greater variety of species. A quick back-of-the-envelope calculation using the *Plymouth marine fauna* indicates we have nearly a thousand species of animal alone recorded from this beach.

By and large, as the size of an area increases so too does the number of species found in that area. This is referred to by the technical, but deceptively simple, term 'the species-area relationship'. And this holds not just for Wembury but for other beaches and for woodlands, grasslands, freshwater streams, whole countries, islands and even regions and continents – just about whatever environment, and whatever scale, you can think of. Most of our studies are based only on animals or plants. However, Claire Horner-Devine and her colleagues recently demonstrated for the first time that bacteria also show a species-area relationship over a scale of centimetres to hundreds of metres in a New England salt marsh (2004: *Nature*, 432,750). Even more recently, Thomas Bell and his co-workers found that the numbers of different bacteria were greater on big islands than small ones (2005: *Science*,

308,1884). The species-area relationship is one of the strongest relationships we observe in biodiversity studies. To many, I suppose this relationship seems intuitive – but if you think it through, it doesn't necessarily have to be that way. Why must animal, or plant, or microbial species be unevenly distributed in space? In fact there is still no real consensus among scientists as to why we should get such a relationship. There are a number of different ideas and few of them are mutually exclusive.

There are two ways in which species can be added to any particular area. They may come in from other areas as immigrants, or else the species already present for some reason start to throw off new species, to speciate (we'll come back to this). And there are two corresponding ways in which species can 'disappear' from that area. They may leave (emigration) or they may go extinct. Put simply, the total number of species in an area must depend *to some extent* on the relative importance of each of these four features and their interactions and so it is here we begin our search for the explanation(s) underlying the species-area relationship.

Current opinion holds that when we are dealing with very large areas, much larger than our Wembury Bay and the surrounding area, on the whole immigration and emigration become less influential. If this is true then it is the balance between speciation and extinction that is critical in determining the relationship.

But equally likely, the species-area relationship may just be the result of larger areas containing a greater number of different types of habitat. Our original one square metre on Wembury beach was almost exclusively open rock. But as the area we were interested in became bigger then not only did we get more open rock to explore but we started to encounter rock pools, crevices, under-stone environments and, finally, sandy shores and cliffs and grassland. These different habitats each contain a whole new set of different species. And what works on a local scale could also take place at a number of different scales – the Plymouth region with a multitude of different habitats and environments, the south-west of England, the UK, north-west Europe and so on. Species number increases with increasing area just because different species live in different types of habitats. Certainly, this was considered to be so for the salt-marsh bacteria referred to above.

There is good reason why the sets of explanations we have considered above could explain the species-area relationship. Then again, there is also the unpleasant possibility that this relationship is not actually a relationship at all. The fact is, the bigger area you choose the more samples you have to take, and the more samples you take the greater chance you have of finding new things.

So, the larger the area the greater the number of species present as a result of: (a) the balance between immigration and emigration, speciation and extinction, with the last two probably being more important in very large areas; (b) the greater number of different habitat types and environments in larger areas. However, over and against this pattern there are a number of fundamental differences between different areas. We will start with what is a truly global-scale difference, that between the oceans and the land.

those who go down to the sea in ships ...

Two-thirds of our planet's surface is covered with water, salty water. Furthermore, three-quarters of the sea bed is abyssal – that is, it occurs at a depth of between 5 km and 11 km, it is in total darkness and is as cold as the inside of your refrigerator. As well as the large surface area of the sea bottom, there is also all that overlying water which is potentially living space for life. This said, it is perhaps surprising that about 98% of all marine animals and plants are bottom-living, with only 2% living in the water mass, and even then restricted mainly to the upper, sunlit layers. And yet in some ways living things in the oceans are more diverse than those on land. Of the hundred or so body plans (phyla) that characterize current life on earth, at least two-thirds of these are restricted to marine environments. There is only one phylum, the velvet worms, Onychophora, which is restricted to the land. And this greater diversity seems to hold even for lower-classification categories. For instance, nine out of ten classes are marine. It's only when you come to the level of species that things are not so straightforward.

Less than one in six of the species described so far are marine. This could mean one of two things. It could be that the number of species in the oceans actually does outnumber those on land and that we end up with such a skew in described species simply because it is so much easier to collect land organisms compared with sea organisms (remember, 75% of the oceans have a depth greater than 4 km). Certainly the case has been made for the deep ocean being more diverse than, say, the same area of rainforest. But this is based almost entirely on scaling up some localized measures of biodiversity, which seems to indicate that deep-sea nematodes are hyperdiverse. Not everyone accepts the assumptions and calculations involved and the debate continues. The alternative view, which is not so hotly contested, is that there really are more species living on land than in the seas, despite the fact that the ball-park species numbers we have at present will probably change.

Whichever way you look at it, we have a major land–sea contrast. While there are a greater number of bodyplans in marine environments, the land has a much greater number of species. Why should this be? Lord Bob May put forward a number of suggestions that might help us. First, as life began in the sea, and has been present in marine environments for a considerably greater period of time than it has on land, the early 'innovations' that gave rise to the major phyla we looked at in the previous chapter all took place here. Second, land habitats are more elaborate and more different from each other than marine habitats, and such environmental diversity is thought to promote biological diversity. Linked to this second point it is believed that plant-eaters in the marine environment tend to eat just about anything, compared with land plant-eaters, that tend to specialize, even to the extent that one animal species will feed off one plant (or even just particular parts of one plant) species. Such specialization may be linked to the throwing off of new species (but, as a word of caution, it is difficult to see which one is cause and which is effect). Finally, land life tends to be on average bigger than sea life. For example, photosynthesis on land is carried out by grasses and trees while in the oceans it's carried out almost entirely by microscopic life. As it's supposed to be easier for smaller things to get around, the idea is that the small things should not throw off new species so readily. All of these are neat ideas, but that is still not the same as actually knowing why there is such a profound difference between marine and land biodiversity.

hotspots: a tale of two definitions

One of the main reasons that the study of marine biology is so popular in the city of Plymouth is the richness of the marine life on the shore and in the coastal waters. This has been a constant draw for scientists and naturalists for more than a hundred years. It is one of the main reasons that I live and work here. It would not be unreasonable in common parlance to say that the Plymouth region is a real hotspot for marine biodiversity.

The fact that biodiversity is unevenly distributed means that there must be highs and lows, hotspots and coldspots. Coldspots could also be recognized at different scales. Some wetland areas have relatively few species, as do some polar regions. The term 'biodiversity hotspot' was introduced by Norman Myers in 1988[www#16]. He designated ten tropical forests as hotspots of biodiversity. However, in the use of the term hotspot Myers meant more than just that they were sites of high species richness. To him they had to have large numbers of species that were found nowhere else (endemics) but he also defined them in terms of how threatened they were. So instead of a 'hotspot' being a term for high species richness, merely describing a natural pattern, from its inception it was wedded to conservation and identifying conservation priorities. So, calling Plymouth a hotspot for marine biodiversity is not necessarily the same as calling one of Myers' forests a hotspot.

Most references to hotspots in the scientific literature follow Myers' usage. In fact, there is now a formal definition of hotspot in that it must be an area which contains at least 1500 species of plant and must have lost at least 70% of its original habitat. Hotspots tend also to be defined using larger organisms such as mammals, birds and plants. Recent (i.e. 2004) inventories of

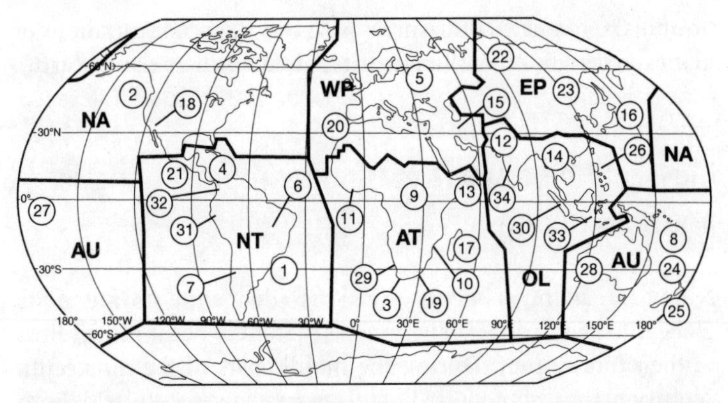

Figure 4 *Biodiversity hotspots (information from* Conservation International, *2005). 1 Atlantic forest; 2 California floristic province; 3 Cape floristic region; 4 Caribbean islands; 5 Caucasus; 6 Cerrado; 7 Chilean winter rainfall Valdivian forests; 8 East Melahesian islands; 9 Eastern Afromontane; 10 Forest coasts of eastern Africa; 11 Guinea forests of west Africa; 12 Himalaya; 13 Horn of Africa; 14 Indo-Burma; 15 Irano-Anatollan; 16 Japan; 17 Madagascar and the Indian Ocean islands; 18 Madrean pine-oak woodlands; 19 Maputaland-Pondoland-Albany; 20 Mediterranean basin; 21 Mesoamerica; 22 Mountains of central Asia; 23 Mountains of south-west China; 24 New Caledonia; 25 New Zealand; 26 Philippines; 27 Polynesia/Micronesia; 28 South-west Australia; 29 Succulent Karoo; 30 Sundaland; 31 Tropical Andes; 32 Tembes, Choco-Magdalena; 33 Wallacea; 34 Western Ghats and Sri Lanka*

Also indicated are the biogeographic regions. AT = Afrotropical/Ethiopic, AU = Australasia, EP = Palearctic (east), NA = Nearctic, NT = Neotropical, OL = Oriental, WP = Palearctic (west)

hotspots identified 25–34 different regions worldwide (Figure 4). While these hotspots once covered about an eighth of our planet's surface, an area roughly the size of Russia and Australia combined, they now cover an area variously estimated at 1.4–2.3% of the earth's surface, an area about the size of India. Exact figures vary according who you listen to, but broadly speaking, half of all plant species occur in these hotspots taken together and about half of all terrestrial vertebrates (72% of all mammals, 86% of all birds, 92% of all amphibians). Madagascar and the Indian Ocean islands hotspots were singled out as areas with very high concentrations of plant and vertebrate families that are found nowhere else on earth.

big-scale biodiversity: biogeographical and political regions

Numerous attempts have been made to divide the surface of the globe into just a couple of large areas that differ naturally in terms of their biodiversity. During the middle part of the nineteenth century naturalists collected plants from all over the world. Both specimens and notes were brought together and analysed in a number of famous herbariums. From such information, in 1866 August Grisebach produced a global distribution map for plants. Andreas Schimper went a little further when he produced a more detailed map in which he grouped the main plant types according to what latitudes – polar, temperate, tropical – they occurred. Studies of plant distribution went hand in hand with, and became intricately linked to, the study of climate, one of the main features believed to be driving plant distribution.

I have on my shelves a beautifully illustrated book entitled *The Geography of Mammals* (1899) by father and son team Phillip and William Sclater (William is first author). They introduce the book by saying: 'Let us ... dismiss from our minds ... the ordinary notions of both physical and political geography, and consider how the earth's surface may be naturally divided into Primary Regions, taking the amount of similarity and dissimilarity of animal life as our sole guide.' They followed a scheme of dividing the globe into six main divisions. This scheme was originally put forward by the father, Phillip, in 1857 in an essay on the distribution of bird species read before the Linnaean Society, and is almost entirely the same scheme that has been used right down to the present day (see Figure 4 where these biogeographic regions are shown on the map of the world, along with the main biodiversity hotspots). Alfred Russel Wallace, who discovered natural selection at the same time as Darwin, came to the conclusion that 'admitting that these six regions are not precisely equal in rank, and that some of them are more isolated than the others, they are in geographical equality, compactness of area, and facility of definition beyond all comparison better than any others which have been suggested'. In Sclater's scheme the land masses are split into six parts, what we now would term biogeographical regions. There are three tropical regions – the Ethiopean region (now the Afrotropics), the Oriental region (now the Indotropics) and the Neotropical region (the Neotropics). These three, however, are home to about 70% of land (as opposed to sea) organisms. And of these three, the Neotropics is by far the most biodiverse.

Such a task of identifying natural divisions in the earth's oceans has not been quite so easy. The reason given for this is that there are not so many physical obstacles to organisms moving around in what is much more a three-dimensional environment than dry land. We know more about the surface of the moon than about the deep oceans which make up so much of the ocean bottom. Thus, it should come as little surprise that there is no list of agreed biogeographical regions for the world's oceans. That said, there are a number of large-scale patterns observable. The Indo-Western Pacific area is believed to have the greatest concentration of marine biodiversity on earth. It also has the greatest diversity of coral reefs and the organisms associated with them. Coral reefs are important for biodiversity generally as it has been estimated that these habitats are home to one in four of all marine species and one in five of all fish species. In contrast with coral reefs, the deep-sea (abyssal) has been considered to be poor in its biodiversity (even allowing for recent claims of it being the most diverse habitat on earth). All of the major invertebrate groups are present in the deep-sea, as well as the major fish groups, but it is the echinoderms (brittle stars and sea cucumbers) – which, if you remember, is not a hugely species-rich phylum – that dominate. Also, while the mass of large marine animals is about 0.2 kilograms for every square metre of shallow sea floor, it falls to 0.2 grams (a thousandfold difference) below a depth of 3 km. On the whole, more animals are found in deep water beneath the highly productive waters of polar regions than in deep waters in less productive temperate regions.

Delineation of biogeographical regions is our attempt to recognize natural barriers and discontinuities in biodiversity. However, most of the existing information has not been collected and managed on such a mega-scale. Instead we have biodiversity inventories collected at the level of individual countries. While it is not hard to see the reasons for this, information collected in this way must be used and presented with some caution. Countries may be delineated on the basis of natural features. However, more often than not (as noted by Phillip and William Sclater above) they are entirely human constructs – look at a map of Africa if you need convincing – with (depending on their size) little or no biological significance. Thus, while biogeographical regions may be the scale at which we would like to understand and manage biodiversity, in reality it is often at the level of individual countries that scientific studies are initiated, decisions made, and conservation measures conceived and implemented. This being so, how is biodiversity distributed between different countries?

It turns out that we think that most countries contain comparatively few species. There are, though, a small number of exceptions. In fact, about a dozen countries are estimated to contain one-half to three-quarters of the world's species. Most of these countries are in tropical regions and are among the poorest on earth, an observation that we will return to. I said we think that most countries contain few species; this is because there are very few countries with anything near a good, never mind complete, list of their resident species.

latitude for life?

Alexander von Humboldt (1769–1859) carried out some detailed investigations on the relationship between plants and their environment in tropical South America. One of his big findings was that changes in the types of plant found as you climbed up mountains was very similar to what you see if you journey from lowland tropical to polar latitudes. Since then others, including Alfred Russel Wallace who co-founded the theory of natural selection

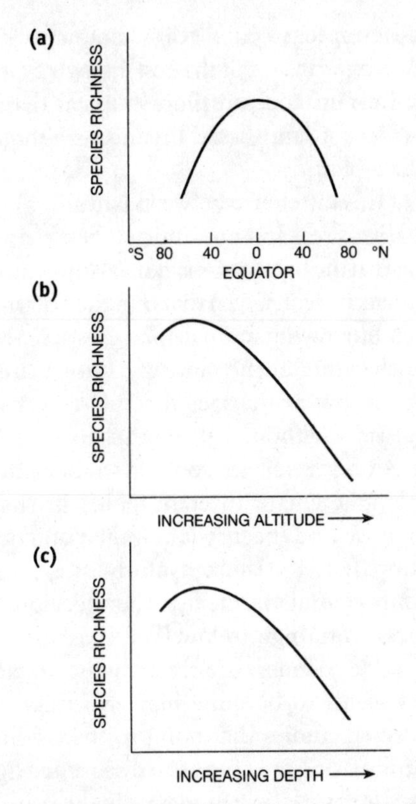

Figure 5 *Three graphs showing the relationships of latitude (a), altitude (b) and depth (c) to species richness*

with Darwin, have observed that the number of species present in an area increases as one goes from temperate (high) latitudes towards the equator (Figure 5a). This appears to be true not just for living organisms but for fossils too. The same pattern even applies if we look at higher units of classification such as genera and families. Interestingly, this increase as one heads towards the equator differs between the northern and southern hemispheres. Species richness increases more sharply in the former compared with the latter, so much so that some people have described the globe with its biodiversity as not so like an orange but a pear shape. And even between species, some species are more 'tropical' than others, with one of the best-known examples being butterflies, which, for instance, are more 'tropical' than birds. This pattern is very clear for land-based organisms although slightly less so for marine organisms.

How much of this difference between latitudinal gradients of marine and land life is real and how much it has to do with problems in sampling marine organisms is not yet known. The pattern for deep-sea animals is clear. Given that our quantitative information for deep-sea burrowing animals, for example, comes from sample areas which in total are no more that one square kilometre, this is amazing. So too is the fact that there is a pattern for planktonic organisms, although it appears to be a little more complicated. It is only when we look at studies attempting to determine if this gradient of diversity holds in shallow-water environments that we find the jury is definitely still out. The evidence is conflicting. True, if we look at information on perhaps the best-studied group, coastal marine fish, they do show an increase in species richness from high to low latitudes, but it cannot be denied that the species richness of animals living on the sea bed in Antarctic waters seems to be quite high. No matter what, the number of published studies that point to there being no such general pattern for marine life cannot be discounted lightly.

If the evidence for latitudinal gradients in diversity (at least in many, if not all environments) is relatively clear the same cannot yet be said for why we see these patterns. Many explanations have been put forward. None of them are mutually exclusive and there are numerous adaptations and refinements of each in the scientific literature. First, it could simply be that there is a greater area of tropical habitats compared with temperate habitats just because of the way our planet is shaped. Related to this 'problem' with the shape of the earth is the possibility that there may be more of the sun's energy reaching, and therefore available in, tropical regions compared with temperate ones. More available energy equals more species. Why it should follow that more energy means more species rather than just more individuals of the existing species is not clear, but there are even greater difficulties with this idea. Basically, when energy availability is measured for different latitudes and matched with species richness sometimes there is a positive relationship, and other times, well, it's complicated and there can even be a hump-shaped relationship. Finally, it has been suggested that tropical animals have been exposed to more unbroken stretches of evolutionary time to throw off new species, fill habitats and so on. In this scenario the tropics were not so dramatically affected by, and the evolutionary process was not interrupted by, such climatic or geological upheavals (e.g. drying out and glaciation) as higher latitudes were. The only slight problem with this idea is that there is little hard evidence that evolutionary rates are greater for tropical organisms. Furthermore, the tropical climate is not as stable over large periods of time as was once believed. There is some evidence emerging that the present high species richness in the tropics is not a slow accumulation of new members over time, but in fact is due to a relatively recent 'outburst', in some cases as little as ten million years ago. No matter what, the latitudinal gradient in diversity must ultimately be produced (like the species-area relationship we met earlier in this chapter) by the balance between the rates of four different processes: speciation, extinction, immigration and emigration. At the large scales we are considering, immigration and emigration are likely to be small players. So in the scientific literature we find the tropics being portrayed either as a 'cradle' (high speciation rates) or a 'museum' (low extinction rates) of diversity. Currently there is good supporting evidence from fossil marine life for the 'cradle' idea, but probably the truth is that latitudinal gradient in diversity is a balance between 'cradle' and 'museum'.

lessons from the tops of scottish mountains

How does diversity change the higher up you go? When much younger I climbed many of the Scottish Munros (mountains over 914 m or 3000 ft) with a bunch of friends from Glasgow. One of the striking features of each climb was that you always set off surrounded by plants and grasses and trees, and an incredible host of insects, rabbits and birds. However, by the time you reached the summit you were invariably on hard rock, with the occasional lump of lichen or moss just clinging on, one or two tough-looking old birds hovering overhead (if you could see them through the torrential rain, mist and cloud) and no midges! It wasn't a quantitative survey but I would have said that biodiversity definitely decreased as you ascended. And while I love the Scottish mountains passionately, they do not really compare in terms of height or scale to, say, the Andes, the Rockies or the Himalayas. And yet a mere 914 metre climb in Scotland can easily result in a reduction in temperature roughly equivalent to the temperature difference between Wembury beach and Madrid in Spain.

So when we trawl through the quantitative published studies, do we see a reduction in biodiversity with increasing altitude? As early as the 1800s Humboldt had noted that the number and type of plant species changed as he ascended the volcano Chimborazo in the Ecuadorian Andes. And taking together all the quantitative studies that have been carried out since, the simple answer is yes, there is a relationship – well, of a sort. Whether we're investigating bats in Peru or treehoppers in Columbia, the number of species decreases with an increase in elevation. However, for flowering plants in the Kingdom of Nepal or ants in Colorado, to take two examples, the pattern is a little more complicated. They both show a sort of hump-shaped relationship (Figure 5b). That is, there is at first an increase in diversity with altitude. Then, as you continue to ascend, the pattern is reversed and you get a decrease in diversity with altitude, but the number of species at the bottom of the mountain is always much larger than the number at the top. Now, given what we've already covered on species-area relationships and the mechanisms underpinning latitudinal gradients in diversity, it may have already occurred to you that different altitudes don't all possess equivalent surface areas. So maybe the relationship we see with altitude is affected, or even determined, by the fact that habitat area almost invariably decreases with increasing altitude? A study of tropical South American land birds found that species richness decreased with altitude. But what happens if you take altitude-related differences in habitat area into account in your analysis? Using the same information but this time standardized for area, the relationship was altered slightly. Instead of a decline in species richness with altitude there was a hump-shaped relationship. It has been suggested that the underlying reason for such hump-shaped relationships lies in an idea we've met before – energy availability. The theory goes that energy availability may be greatest at intermediate altitudes. It'll be interesting to see how and when this theory is tested.

Another theory put forward to explain the decrease in species richness with increasing altitude (although it doesn't really explain 'the hump') has to do with the isolation of mountain tops and similar heights. Travel between mountain tops is difficult unless you can fly. Thus emigration and immigration are low. Such isolation seems to result in new species being formed but at the same time a much greater possibility of populations going extinct. So, the argument goes, the balance between these two processes results in low species richness and the presence of many local (endemic) species at altitude. Why extinction rates should be greater with isolation is not all that clear, although it has been pointed out that ecological processes which are independent of density are more noticeable at higher altitudes, and this could possibly lead to greater extinction rates. While the pattern is becoming clearer, as so much with these big scale patterns we still have a lot to learn about the mechanisms that underpin them.

out of your depth

In perhaps the first book ever fully devoted to marine studies, *Histoire Physique de la Mer* (1725), Count Luigi Ferdinando Marsigli recorded his investigation of animals living in the 'deeps' of the Bay of Lyons, France. Similar findings were made for depths of around 25 metres in the Adriatic in 1750. Edward Forbes, a key figure in the study of the sea in the first half of the nineteenth century, from his early twenties spent a considerable amount of time dredging for animals in the seas around Europe. On board H.M. Survey Ship *Beacon*, he dredged to depths of up to 420 metres in the Aegean Sea. Based largely on the results of this voyage he put forward a hypothesis in 1841 that the 'zero of animal life was probably about 300 fathoms [about 550 m], below which extended a lifeless or "azoic" zone'. Despite the fact that Forbes clearly thought he was putting forward a hypothesis to be tested, within a short space of time it had become accepted fact that diversity decreased with depth until a critical, quite shallow depth, below which there was nothing. This belief was relatively short-lived (despite claims to the contrary in some textbooks) as deeper and deeper sampling became possible, and many underwater surveys in connection with laying the first submarine cables found life at all depths. In a book entitled *The North Atlantic Sea-Bed* (1860) there is a drawing of a brittlestar brought up from a rope that had been at a depth of 2300 m. The voyage around the globe by HMS *Challenger*, taking measurements of the sea and sea floor and collecting animals from great depths, finally brought the existence of deep-sea life, often bizarre and fascinating, to public attention. There was a feeling that biodiversity decreased with depth but such a relationship was not based on quantitative information and analysis. There were, however, still creatures in the deepest depths.

It is only relatively recently that we have been able to put together information allowing us to assess scientifically any relationship between depth and biodiversity. Even then the data are scarce. For many groups, both invertebrates and vertebrates, species richness decreases with increasing depth. For marine 'woodlice' living in the northern seas, sixty-five species were found in relatively shallow water. As you go deeper there is a progressive decline until four kilometres down there are fewer than ten species present. The same sort of pattern holds for fish species, which halve in the number of species present over a two-kilometre depth range. However, we also see the hump-shaped distribution we encountered when discussing altitude. For organisms free-floating in the water column, the peak in the hump is between 1–1.5 km deep, 1–2 km deep for big animals living on the sea bed, and 2–3 km deep for large animals actually burrowing in the sea bed (Figure 5c). Interestingly, in this case the size of the particles that make up the sea bottom may be more influential in determining species richness than the depth those sediments occur at. If allowance is made for different sediment types, it is not uncommon to lose any relationship that has been established between diversity and depth. So our current understanding is that the relationship between diversity and depth is perhaps a little more complicated than the initial belief that numbers of species decreased with increasing depth.

staying close to home

Not all species are found everywhere. There are a few, like the moon jellyfish *Aurelia aurita*, which appear to have a worldwide distribution (although very recent genetic analysis indicates that there could be as many as nine different 'types'). In fact, most species appear to be very restricted in where they are found. Such a species is called an endemic and the restriction in where it occurs is referred to as endemism. We've already encountered endemism when we discussed hotspots and coldspots of biodiversity above, and found that the number of endemics was used to help identify hotspots. At one extreme there are organisms which occur only in one lake, or on one mountain. For example, Crompton's Orcutt grass is found only in one vernal pool (small pools that fill during the winter rains and dry during the spring to become filled with flowers) in the Jepson Prairie of Solano County, California, USA. In fact, all five species of Orcutt grasses depend upon vernal pool habitats in California for their continued existence, and they are found nowhere else. They contribute to the 30% of Californian plants that are found only in California. Compare this with 1% for the United Kingdom, which has an area about three-quarters that of California. Generally, some very small areas can have a disproportionately large number of endemics – oceanic islands, such as Hawaii and the Galapagos, for example. Isolation, whether it's vernal pools, islands or mountain tops, is firmly linked with endemism, although there is also a distinct pattern where the number of endemic species in a given area tends to increase as one heads from temperate to tropical latitudes.

Despite everything that's gone before, it would be wrong to give the impression that the range size of an endemic is always small. For example, turkeys, when living free and not frozen on supermarket shelves, are confined to the Neoarctic biogeographic region, a not insubstantial area. And four out of five Australian plants and animals are endemic, creatures such as the koala and the red kangaroo, because of the isolation of the continent from Asia for tens of millions of years.

congruence: the holy grail of biodiversity?

It should now be apparent that putting together a complete atlas of biodiversity would be a colossal, not to say impossible, task. In fact, even trying to work with indices of biodiversity, such as species richness, and documenting patterns related to latitude, altitude and depth is in itself a huge undertaking. Much of our information is based on plant or animal groups, and when we refer to animals there is a strong bias towards (not surprisingly) the conspicuous groups – the birds and mammals. We know very little about the invertebrates and much, much less about microscopic life in all its forms. And yet even with the disparate information we have, common themes are beginning to emerge.[www#17] As we have seen, with all the appropriate caveats and provisos, species richness does decrease for a reasonable number of quite different plant and animal groups as you: (a) head away from the tropics; (b) ascend to great heights up mountains; and (c) descend to great depths in the oceans.

Given such patterns, it is not unreasonable to believe that what is happening with one group of organisms could be very similar to what is happening with another group of totally unrelated organisms. Just as the human pulse can be used to tell you so much about the normal workings of the human body because of the interrelatedness of the different body systems, could it be possible that patterns for one (or a number of) group(s) could tell you about other groups, including some that we may never have time to examine? This is what is referred to as congruence. Congruence would make possible our atlas of biodiversity and we could even include groups of organisms for which, because of time and financial constraints, we have no original information. And here indeed is the holy grail of biodiversity. Think of the huge implications of successful congruence for those responsible for monitoring the environment and its associated biology. It could alter beyond recognition how we construct, introduce and enforce conservation measures and how we 'use' biodiversity resources.

How far have we got? Attempts have been made to search for congruence between groups for which we already have reasonably good information. The results of these studies are not always encouraging, but it is early days yet. Back-of-the-envelope-type analyses seem to highlight degrees of congruence between different groups, but often when the detailed study is carried out the outcomes are much more mixed: some groups 'yes', some groups 'no', some groups 'don't know'. Perhaps for the moment attempts on the largest scale possible are the best we can manage. In 1997 Paul Williams, Kevin Gaston and Christopher Humphries published a study in the *Proceedings of the Royal Society* examining the worldwide distribution of numbers (at the level of family, not species) of seed plants and tetrapods (four-legged animals with backbones – mammals, reptiles and amphibians). These are groups for which we have some of the best worldwide distribution information. Interestingly, there was a fair degree of congruence. On a map of the world coloured according to the total numbers of families of these groups found at that location it was possible to see previously identified hotspots of biodiversity – in Columbia, Nicaragua and Malaysia. Also most satisfying was the gradient of colour on either side of the equator, indicating that there was a latitudinal gradient in diversity exactly as we saw earlier in this chapter.

So overall, we are beginning to put together some of the basic patterns of biodiversity. What is also clear, however, is that even building these patterns requires a lot more work; we are still only beginning to uncover the reasons why such patterns exist, and we have a long way to go before we can use well-known groups to predict things about less well-known groups. There is an ancient map of the northern Atlantic drawn by Sigurdur Stefansson in

Iceland around 1590. What are now the British Isles, Iceland and Norway are instantly recognizable. Gronlandia, which presumably is Greenland, and Helleland (Baffin Island) are pictured as promontories of a large continent in the west. A number of the other locations in the Arctic seas and on this great western continent are much more difficult to place. Compared with the most recent edition of *The Times Atlas*, which covers the north Atlantic, Stefansson's map looks a little comical, basic in the extreme. And yet that map and its predecessors were used in some of the greatest feats of exploration carried out by the Scandinavian countries. Great things were accomplished even with such a basic map. The 'atlas of biodiversity' we have at the present time resembles Stefansson's map. It is basic. It is provisional. But you have to start somewhere. And as was the case with Stefansson's map, there is no reason why it cannot be used to plan and to act upon. And if our knowledge of present-day biodiversity is incomplete and provisional, what of the subject we turn to next: the origins of that biodiversity from back in deep time right through to the recent past?

chapter four

diversity, extinction and deep time

There had always been very simple flowers on the little prince's planet, with a single ring of petals, occupying very little space ... [b]ut one day, from a seed blown from no one knew where, a new flower had come up; and the little prince had watched very closely over the small shoot which was not at all like any of the other shoots on his planet.

The Little Prince, Antoine de Saint Exupéry

one every twenty minutes?

A colleague of mine, Paul Ramsay, spends a lot of time in the high Ecuadorian Andes. He is interested in describing and preserving the biodiversity of these incredibly beautiful highland areas. One morning over coffee he showed me a picture of a plant in flower. It was brand new to science and had just been described from specimens found only at this one location beside a high alpine lake, Lake Luspa, at 3900 m. The plant is called *Loricaria cinerea.* His tone changed as he told me that the area had recently been cleared by burning, and all of the plants destroyed. There was a moment of quiet before we continued on to another subject. Just newly discovered, *Loricaria cinerea* was now extinct – lost for ever.

Three species disappear every hour, according to E. O. Wilson, an entomologist and populist, whom many see as the founder of modern-day interest in biodiversity.[www#18] Species go extinct. Sometimes species disappear from a local area, but can still be found in other parts of the globe (local extinction). For others, like *Loricaria cinerea*, extinction is, as far as we know, global. The mass of biological information carried by the genes disappears forever. But extinction is not new. And if we look back into deep time, at the history of past life on earth, we will see that the appearance and disappearance of species played a major part in forming present-day biodiversity. Perhaps learning about the history of biodiversity, and the role of extinction in that history, will help us to see present-day extinctions in a clearer light. So in this chapter we will briefly look at the history of life on earth from its beginning to the present-day, paying particular attention to the key events, including extinction patterns, that helped to shape that history. At the end of this chapter and into the one that follows we will look more closely at the extent and magnitude of the effects that humans and their relatives have had on biodiversity throughout the past quarter of a million years and how today's biodiversity crisis compares with 'natural' extinctions in deep time. So to start at the beginning – quite literally, that is.

a year in the life of biodiversity

But before diving straight in to the major patterns of ancient life on earth, 4600 million years is a lot of years to comprehend. As humans we find it difficult enough to think in terms of hundreds of years, never mind thousands of millions. So, as we tried in chapter 2 to get some perspective on huge distances, we need to do the same for deep time. The writer of the first book in the Judeo-Christian scriptures, the book of Genesis, contracted the creation of everything to six days, with biodiversity being produced on the last three. Quite a neat way of teaching – taking your reader from the unknown or inconceivable to the known by means of something analogous, more everyday and much easier to grasp. At least that was the view of St Augustine of Hippo (AD 354–430). While drawing lessons on God (arguably the main purpose of the text, whatever the creationists claim) and on life from the structure and content of the first chapter of Genesis, Augustine actually believed that creation took place in an instant. Calendars are devices we work with all the time and most of us are reasonably good at thinking on this timescale. So what would the history of biodiversity look like in broad brush strokes if we contracted our 4600-million-year history to occupy the same time as our calendar year?

The world begins very early on 1 January. In terms of clear life signs (i.e. definite fossils) we have to wait until the end of March before the first prokaryote (remember them?) cells appear. We have an even longer wait, through spring to the end of the summer, for the first eukaryotic cells to appear at the end of August. Most of the year is over before we get the first definite animals midway through November. Up until this point there has been no life on land. The first invaders, plants and arthropods get there at the very beginning of December. Something really nasty happens in the lead-up to Christmas and nine out of ten species become extinct. And yet within a day life is up and rolling again. Many forms have disappeared, but others have taken their place. It is now the age – or, should I say, the two weeks – of the ruling reptiles including the dinosaurs. These creatures dominate the air, sea and skies until just after Christmas, when they disappear. In the few remaining hours of the year on Hogmanay humankind appears on the scene. With a couple of ups and downs through the year overall biodiversity has increased dramatically from the end of March to New Year's Eve. As for recorded history, it occupies the same amount of time as it takes you to count one, two, three, four, five, six seconds. Most of the history of biodiversity is solely the history of microbes, at least until mid-November, when it becomes the history of lots of microbes and some plants and animals (see Table 1, p. 75).

precambrian – before life?

We encourage our children to go to school so that they can learn and go on to achieve great things. Perversely it was 'dogging off' (i.e. AWOL) that led a schoolboy, Roger Mason, to an incredible biodiversity discovery. While playing in Charnwood Forest, Leicestershire, UK, he found the impression of what looked like a feather on the rocks there. He told his dad and the pair approached some geologists at the nearby university concerning the boy's find. In the words of one of those geologists, he 'quite frankly did not believe the boy'. After all, the rocks in Charnwood Forest belong to what geologists refer to as the Precambrian, and at that time there were known to be very few fossils in rocks older than the Cambrian (544 million years ago). In fact, before the second half of the twentieth century, the Precambrian period was still thought of by its older name, the Azoic (without life) period. However, going to see the rocks for themselves, the geologists found the fossil of a sea pen, a creature still found half-buried in shallow marine waters today, and looking very much like a thickened feather. They named it *Charnia masoni* after the schoolboy.

Since then the number of Precambrian fossils found has increased dramatically. As we've already seen, the oldest definite fossil is of a prokaryote, contained in rocks which are a staggering 3500 million years old. The oldest definite eukaryotic cell is found in rocks 1500 million years old. In fact, all of the fossils more than 700 million years old are attributable to microbial life. There were claims in 1998 in the journal *Science* that fossilized worm burrows more than 100 million years old had been found, but this has not really been accepted by the scientific community.

To date, the earliest animals that we know about come from rocks that are about 600 million years old, Precambrian rocks (sometimes referred to as Vendian) from Australia, Canada, England, Namibia and Russia. The discovery of this Precambrian life was made initially in Ediacara, Australia. The strange, almost plant-like shapes of many of the animal impressions in these rocks (Figure 6) led to this window into the past being referred to as the 'Garden of Ediacara'.[www#19] When first collected, the impressions were interpreted in the light of the different body plans (phyla) that exist today. There were organisms that looked like jellyfish, and Mason's fossil belonged to the same group as the jellyfish and sea anemones. However, many of them were also quite 'other'. All

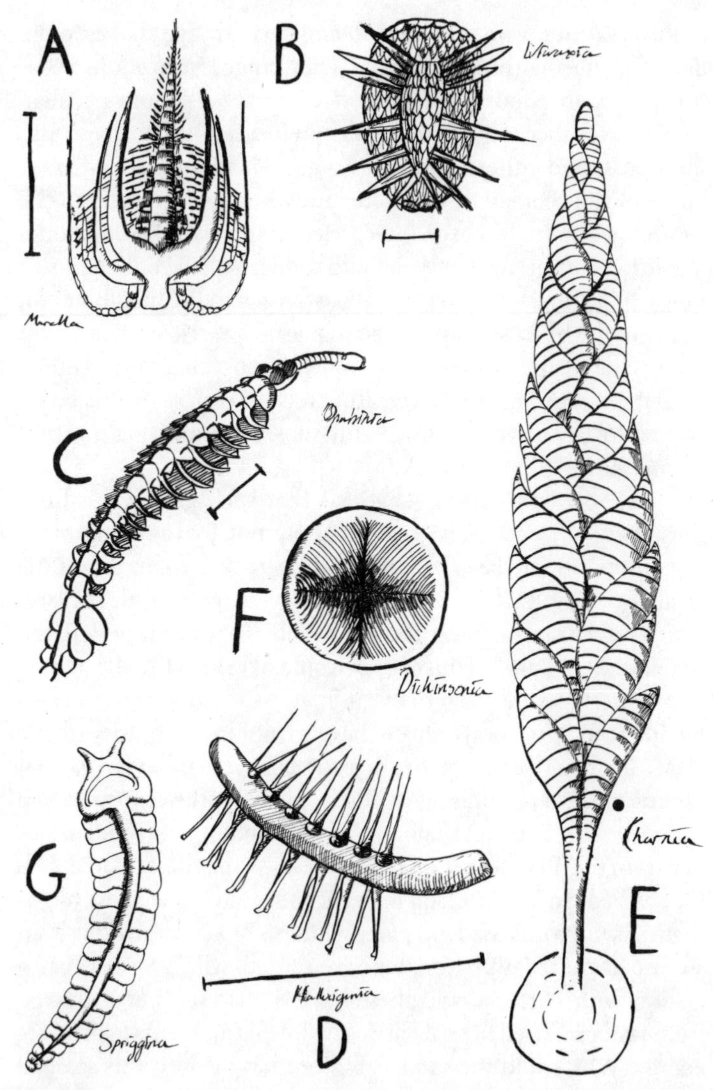

Figure 6 *Examples of some Burgess Shale and Ediacaran animals: A.* Marella, *B.* Wiwaxia, *C.* Opabalina, *D.* Hallucigenia, *E.* Charnia, *F.* Dickinsonia, *G.* Spriggina

Source: Redrawn from various sources by Ben Spicer.

of the creatures were soft-bodied, and very strange shapes, often flattened sheetlike or even leaflike. They ranged from about a centimetre up to about a metre in size. What was more peculiar, though, was that initially there was no indication that these animals possessed either a mouth or a gut. Unfortunately, what we know of the biology of these creatures is only speculation. The strange geometry has been interpreted in terms of the difficulties in getting oxygen into an animal that doesn't have either gills or lungs, or a circulation system to take oxygen to the tissues once it gets into the body. It is estimated that atmospheric oxygen at this time in the earth's history was very low, at most about one tenth of what it is today. The flatter a creature was the easier it would be for oxygen to get to even the deepest tissues. Thus one might expect animals to be thin and leaflike.

In 1986 McMenamin suggested that such gutless creatures must have secured their energy supply not by ingesting other organisms but by keeping within their tissues many tiny little algae. According to this view the algae were responsible for capturing and transforming the sun's light (photosynthesis) into usable energy. Some of this energy would have kept the algae going and the rest was supplied to the animal tissue. So they would be a bit like modern corals which have symbionts, photosynthetic algae, in their tissues. A major problem with this view is that photosynthesis requires light. Today, photosynthesis is restricted to algae either living in shallow waters or floating near the surface as part of the plankton. It is not a process normally associated with the sea bed. So in 1989 Dolf Seilacher suggested that these hypothetical symbionts used not *photo*synthesis but *chemo*synthesis. In some parts of today's ocean there are animals which form an association with bacteria, and these bacteria use sulphur extracted from water released from deep-sea hydrothermal vents to produce energy. Why could the same thing not have driven this sea bed ecosystem 600 million years ago? It's a possibility, but we still don't really know how Ediacaran animals worked.

One of the key questions that emerges from careful examination of the Ediacaran animals is: could such strange forms really be attributable to animal designs (phyla, *bauplans*) that exist today? Certainly some scientists, including the palaeontologist Martin Glaessner, have tried to interpret many of these rock impressions in terms of the phylum Cnidaria (which contains the sea anemones and jellyfish) and the phylum Annelida (the 'true' worms, like the ragworm, the earthworm or the leech). There was even a suggestion that Ediacaran creatures were not animals at all, but were instead large single-celled 'microbes'. Others, such as Seilacher in the 1980s, thought that most Ediacaran animals were animals but could not be assigned to modern-day phyla; they were, in fact, a very early 'failed experiment' in life – a bunch of 'totally other' body designs that never made it through to the Cambrian period 550 million years ago, never mind to the present day. If so, they are creatures that left no descendants and so tell us little about the rise of present-day biodiversity. To be honest, the jury is still out on what they are and how they worked, but the verdict may well have a bit of the 'unsuccessful experiment' view about it.

The bottom line is that we know almost nothing of bacterial, fungal or plant fossil biodiversity from its origins in deep time until comparatively recently. Undoubtedly, biodiversity increased from its inception until the end of the Precambrian 544 million years ago. But the patterns are unknown. What we can say is that microbial life, bacteria and algae dominated throughout this time, and so, because of the huge timescale involved, have dominated biodiversity for most of its history.

'explosive' cambrian

In sedimentary rocks that are about 544 million years old, irrespective of where they are found in the world, we see the 'sudden' appearance of many different kinds of fossils. They are the remains and impressions of all the major animal phyla still present on earth today, as well as a few phyla that do not seem to have left descendants. Today animal life is considerably more diverse than in any other of life's kingdoms and much of that diversity appears around 544–505 million years ago in the Cambrian (named after the Cambrian mountains in Wales) period. The fossils are predominantly forms with hard parts, shells and armour. This is the first appearance of such forms and the volume and variety is so impressive that this appearance in the fossil record is often referred to as the 'Cambrian explosion'. The very earliest Cambrian rocks from places as far flung as Siberia, China, India and Canada contain a wealth of small (few greater than 1 cm in diameter), difficult-to-identify shells or skeletons, the so-called small shelly fossils or SSFs. And within a very short space of time we find comparatively large animals appearing in the fossil record. Over what appears to be a very short time in geological terms we have life, and life in abundance. As yet, all of this life is concentrated in the seas and oceans. There are no land animals.

When we looked briefly at some of the key characteristics of the most species-rich groups in chapter 2, we had to miss out a large number of even major phyla. So, as we introduce each of the major animal phyla that make their appearance in Cambrian rocks, if we haven't encountered them before, we'll spend a little time looking at some additional detail of the modern-day forms. (Remember, all of the phyla referred to in this book are listed in the Appendix.)

As is still the case today, the joint-legged animals, the arthropods, dominated Cambrian seas. But the dominant forms, the phylum Trilobita, are not creatures we would immediately recognize, mainly because trilobites have been extinct for a very long time. They look a bit like giant woodlice. In fact trilobites are more closely related to scorpions and spiders than crabs or woodlice, and although they retained that basic woodlouse shape they displayed a tremendous variety of forms, often showing different degrees of 'spinyness'. Crustaceans, too, make their first appearance in Cambrian rocks. While definitely sharing the crustacean *Bauplan*, they look quite different from the shrimps, crabs and lobsters of today.

Another dominant, but unfamiliar, group are the graptolites. More like bizarre etchings than animal impressions, these twig-like creatures appeared midway through the Cambrian. They were colonial animals living in connected tubes. Graptolites are now extinct (mostly by the end of the Devonian although some held on into the Carboniferous), but they are related to what is now a very small group with only 100 living species, phylum Hemichordata. There are a couple of shelled species that still resemble graptolites although modern-day forms are mostly wormlike (acorn worms) and look nothing like their ancestors. Adult acorn worms, *Glossobalanus sarniensis*, are very rarely found in the Plymouth area. They must be relatively common, though, as the larvae can at times be very abundant in plankton. *Glossobalanus* looks nothing like a graptolite.

More familiar is the phylum Mollusca, snails and armoured cuttlefish. We've already encountered them in chapter 2 as the eighth most species-rich group. Some of the earliest molluscs look a little like periwinkles that we get today, with one of the most common species, *Helcionella*, more closely resembling a limpet. The armoured cuttlefish is basically a soft-bodied form living inside a small curving shell. Clams and their relatives do not really get going until after the Cambrian.

Brachiopods look very much like clams on stalks. But they are quite different inside. Most of their body is made up from a multi-coiled, intricate, often beautiful feeding organ, the lophophore. Although they were big news in the Cambrian and later – there are 12,000 fossil species – today the phylum Brachiopoda is made up of a mere 300, mainly deep-water species. Brachiopods are now but a shadow of their former selves in terms of their contribution to biodiversity. This said, they can still be extremely abundant. There are some beaches on the Namibian coast, running for miles, which are made up almost entirely of dead brachiopod shells. Interestingly the lamp shell, *Lingula*, which lived in the intertidal zone of Cambrian beaches, can still be found, essentially unchanged and living in the same sort of habitat in New Zealand 525 million years later.

Until relatively recently it was thought that vertebrates, back-boned animals, belonging to the phylum Chordata (which they share with such unlikely bedfellows as the sea squirts and what has been described as a fish without fins, the lancelet) did not originate until after the Cambrian. Finds of fish scales in Cambrian rocks have shown, however, that this was definitely not the case. Certainly, there are also fossil lancelets in Cambrian rocks from

China and Canada, which shows that chordates appeared relatively early on the scene.

There is one phylum of reef-building creatures, similar to sponges, that were abundant at the beginning of the Cambrian but never made it to the end – the Archaeocyatha. They have the dubious distinction of being the only phylum that, by common agreement, has gone extinct. There were no sea urchins, sea lilies, brittlestars or starfish in early Cambrian seas, although there were in abundance creatures called Edrioasteroideans (now extinct) which did a good sea urchin impression. There are also fossilized remains of some of the more soft-bodied phyla and we will come to those shortly.

One of the key questions when considering the rise of (at least animal) biodiversity in the Cambrian is: why now? What was so special about Cambrian times that allowed the establishment of all the major groups we see today? Before this time all of what we would recognize as present-day continents were actually bunched together to make one supercontinent. Just before and during the Cambrian this land mass was breaking up, increasing the area of shallow seas around the continents, creating new habitats. It is in shallow seas that the Cambrian animals lived, as does much of present-day marine biodiversity. It is thought that the dramatic increase in shallow sea areas ripe for colonization was accompanied by an increase in diversity. Some scientists have also suggested that the Cambrian was much warmer than today and/or that oxygen levels increased (due to an increase in the number and diversity of plant and algae photosynthesizing) to a critical level, allowing the construction of relatively large animals. Yet others have drawn attention to the fact that much of the genetic machinery, *Hox* genes and the like, seem to have evolved around this time, and the major genetic control systems of all animals originated and became relatively fixed in the Cambrian – there are no new (major) phyla after the Cambrian period. Ideas are plentiful but we still don't know for sure what kicked off this huge diversification event.

It should be said that not everyone thinks that the Cambrian explosion was the big bang that it is traditionally made out to be. It has been claimed that there is a large, but for some reason unrecorded in the rocks, Precambrian history of many of the major groups. For instance, in 1997 Fortey, Briggs and Wills studied how Cambrian and recent arthropods were related to one another. They found that large tracts of time before the Cambrian were required to make sense of these relationships. For example, the ancestors of the millipedes and insects are believed to have diverged from other arthropods before the crustaceans diverged from the chelicerates and trilobites. If this is true, as the crustacean divergence had occurred by the beginning of the Cambrian, this must mean that there is a long history of the millipede and insect ancestors that we know nothing about. More of a long fuse than a big bang?

Given that the Cambrian is key to our understanding of present-day biodiversity (at least at the level of basic body plans, the phyla), it would be invaluable to see what the sea floor with all its animals actually looked like. Most fossil assemblages are higgledy-piggledy, jumbled-up messes. But there is one set of Cambrian fossils that show excellent preservation. They are believed by some to be so crucial to how we understand that biodiversity developed on earth, that this group of fossils, their discovery and investigation formed the basis of a popular science book entitled *Wonderful Life* (by palaeontologist and popular science writer Stephen J. Gould). These fossils are collectively referred to as the Burgess Shale.[www#20]

The Burgess Shale was discovered in the first part of the twentieth century by an American geologist, Charles Walcott, secretary of the Smithsonian Institute. He collected around 65,000 specimens during a decade of summer trips to a shale quarry (60 m long and 2.5 m deep) on the side of Mount Stephen in eastern British Columbia, Canada. Walcott interpreted the fossil animals he found against the backdrop of familiar body plans, of living phyla. If it looked like a trilobite then it was a trilobite, and if it looked like a worm then it belonged to one of the known worm phyla, in much the same way as Glaessner had done when trying to classify the strange creatures he encountered in the Ediacaran fauna. In the 1960s a palaeontologist from Cambridge, Harry Whittington, and some of his students, re-examined, and to some extent rediscovered, the Burgess Shale animals (Figure 6). What was so amazing was the excellent preservation of the specimens. Even soft-bodied forms, and lots of them, were preserved in exquisite detail. While many of the forms were familiar, there was much that it was hard to pigeonhole into existing phyla.

Remarkably, just as was the case locally for Wembury beach and globally for present-day biodiversity, the dominant group in the Burgess Shale were the arthropods. And of those arthropods the most common form was a less than 3 cm long trilobite-lookalike *Marella*. Lookalike because, like many of the other arthropods, *Marella* did not exactly fit the *Bauplan*. This is a common theme of the Burgess Shale. The 120 or so species we find in the Burgess Shale allow us to look at the patterning of life, while the pattern is still being woven. The largest animal found was the half-metre long predator *Anomalocaris*, which would have moved through the water by the beating of giant flaps on its sides. Worms were present, as were sponges, soft-bodied forms that do not normally fossilize. And then there were some pretty peculiar forms. It is difficult to know what to make of the aptly named *Hallucigenia*, for example, a caterpillar-shaped creature with seven sets of limbs and seven spines along its back. In all likelihood this creature was probably a distant cousin of current-day velvet worms, so not as strange as it might at first appear to be. These legged worms have a phylum all to themselves, the Onychophora, and with a grand total of about eighty living species they are now land creatures (as already said in the last chapter, the only exclusively land phylum) living in tropical regions. And then there's *Wiwaxia*, looking so much like a half walnut with a load of spines projecting out from it. What is interesting is that the Burgess Shale is typical of Cambrian times, it is not an exception.

Gould described the Burgess Shale as the most important fossil deposit ever found. His book retells beautifully the discovery of the Burgess Shale and the rediscovery by Whittington and his colleagues. Gould's main message in the book was that the Cambrian seas threw up many, many animal designs – far more than we see today. Very few of those designs made it through the Cambrian to form present-day biodiversity. He uses the story of the Jimmy Stewart film *It's a Wonderful Life* (where all-round good guy wishes he had never been born and sees a radically different world without his influence) as a vehicle for his message that if we were to rerun the 'tape of life' then everything would be different. There would be no guarantee that the success stories from the Cambrian would be the same success stories from the Cambrian rerun. Gould's book is eminently readable and because of that is very convincing. However, one of Whittington's Ph.D. students who worked with him on the Burgess Shale, Simon Conway-Morris, wrote a response to Gould in a popular book, *The Crucible of Creation.* In this book he argues that, while there is much that is novel about the Burgess Shale animals, Gould has made too much of that novelty. He also asserts that if you did rerun the 'tape of life' the result would largely be the same; the best designs would always win through. Arthropods were successful and would be successful again on a rerun, not because they were lucky but because they had hit upon a good design. Conway-Morris has extended his ideas on the inevitability of the best designs winning out in his latest, and probably most controversial book, *Life's solution: Inevitable humans in a lonely universe.*

To conclude, we can safely say that at the beginning of the Cambrian we see a huge increase in the types of design that characterized early animal life. All of the major groups we have today make their first appearance. By the end of the Cambrian there was a stabilization of many, but not all, of these new groups. That's not to say that biodiversity stabilized, as we shall see next.

post-cambrian: tinkering with successful designs?

In one sense all of the major innovations in body plan and design had taken place by the end of the Cambrian. The rest was tinkering ... but fairly impressive tinkering, as we shall see.

Palaeontologist Jack Sepkoski, who started collecting dinosaur bones and fossils when he was ten, spent his academic life putting

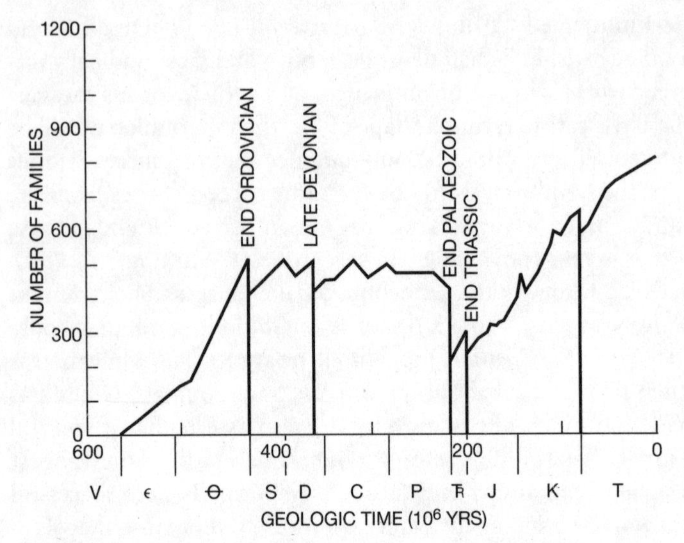

Figure 7 *Sepkoski's graph of changes in marine biodiversity over the past 600 million years*
Source: Adapted from Sepkoski (1992).

together the ups and downs of biodiversity over the past 600 million years. One of the graphs he produced has become almost iconic for those studying how biodiversity has developed. It plots the number of marine animals (or at least the number of families of animals without backbones) against geological time (Figure 7). The pattern produced has come in for some criticism, and still does, but it is probably the best attempt yet. This is against the backdrop of very poor preservation of groups in the fossil record.

In some ways, working out how the major patterns of biodiversity change with time from the fossil record is like trying to reconstruct the story of Shakespeare's play *Macbeth* with access only to Acts II:2, III:4 and V:8. It's just about possible, but you lose a lot of the detail. The problem, or major insight (depending on your point of view), that this analogy highlights is that if you were left with only Acts II:3, IV:2 and V:6 you would have the story of a drunken comedian with bad timing, performing his act just as a murder is discovered, followed by an unrelated woman (whose husband has just dumped her) watching her kid getting butchered – and it all ends in a big fight. The good news is that we also have graphs for some marine protists, land plants, insects and four-legged vertebrates (amphibians, reptiles, mammals) and where they overlap with Sepkoski's graph they show broadly the same pattern. So we may not have much information but what we do have, by and large, corroborates Sepkoski's big picture.

From Sepkoski's graph we can infer that after the initial diversification in the Cambrian we lost a few families (small dip on the graph) before there was an even greater diversification event in the Ordovician period (505–440 million years ago). After the Ordovician ended the number of families seemed to be stable – with bumps – for about a quarter of a billion years. That is not to say there was an absence of activity. The total number of people walking down a busy street may stay constant for some hours, but it is not the same people all of the time. In the same way, families present at the end of the Ordovician were a completely different set from those a quarter of a billion years later. It's just that new families were being added at the same time, and in the same proportions, as old ones disappeared. And this apparent stability also hid the fact that there were major innovations within the basic body plans that appeared back in the Cambrian. Cambrian fish were jawless forms. Now they give way to jawed forms, sporting a number of different fins, that we would recognize as the sharks and rays (jawed fish with cartilage skeleton and no swim bladder to keep them buoyant) and the bony fish. Fish diversity increased through the Silurian period (440–410 million years ago) so much so that the following period, the Devonian (410–360 million years ago), is often, rightly, referred to as the age of fish.

Sometime in the Silurian period, or maybe just before, some of those bony fish, together with some arthropods and plants, made it on to land and founded the first terrestrial ecosystems. Admittedly, early on this life was concentrated near the shore, but with advances in waterproofing bodies and reproductive cells (spores/seeds) the land was not just colonized but conquered within a reasonably short period of time. Whether animals exploiting a new food source followed plants on to land or animals came to prey on other animals ousted from the relatively crowded shallow seas is still very much debated. We can get an insight into what early land life was like by travelling to the small town of Rhynie in Scotland and examining chert rock that is found there. The rock encapsulates, both in 3D and in exquisite microscopic detail, the plants and animals that formed these earliest of terrestrial ecosystems.[www#21] The Age of Fish also saw the appearance of the first four-legged vertebrates, the amphibians and the first insects. By the Carboniferous period (360–286 million years ago) the land was very green indeed, with huge, very diverse forests. The remnants of this biodiversity have literally fossil-fuelled the world's economy for the past 200 years. With the advent of some very tall 'trees', we see the appearance of flying insects, particularly dragonflies – and some giant (1 m wingspan) dragonflies at that.

The Permian period (286–248 million years ago) was a time of major climatic and geological upheaval. It is also the end of this period of 'stability' and a fairly spectacular end, too. Sepkoski's graph shows that half of all families were lost at the end of the Permian, but this hides the true magnitude of this crash in biodiversity. There was a mass extinction of marine invertebrates, with perhaps as much as 95% of all species disappearing. Some groups, such as the trilobites and sea scorpions (aquatic relatives of spiders), disappeared forever. The change in the character of animal life between the Permian and the period that follows, the Triassic (248–213 million years ago), was so profound that it quite literally marked the end of an era, the Palaeozoic ('first-life') era.

The Mesozoic ('middle-life') era was the time of the ruling reptiles on the land, in the sea and in the air, and a new type of life in the oceans, with an expansion of shell-breaking predators and disruptive sediment movers. There was a marked increase in biodiversity, only slightly less rapid in its ascent than the diversification event we saw in the Ordovician. Sometime in the Triassic period, or perhaps even before that in the Permian, the first mammals appeared – small, furry nocturnal creatures. But it is the ruling reptiles that were in the ascendancy in the Triassic and dominated during the Jurassic (213–145 million years ago). Originating from these reptiles was a flying dinosaur that gave rise to the birds. The

Cretaceous period (145–65 million years ago) witnessed the origin of the first flowering plants, the angiosperms.

The rise in biodiversity over the Mesozoic suffered a small setback at the end of the Cretaceous period. The setback was large enough to be seen as the end of this era and the beginning of the Cenozoic (modern-life) era. This is because many of the ruling reptile groups disappeared for ever, as did numerous invertebrate groups such as the ammonites. But in terms of overall biodiversity, if anything the rate of increase accelerated greatly at the beginning of the Tertiary (65–1.8 million years ago). With the advent of the Tertiary period 65 million years ago, we see a picture of biodiversity which has more in common with the present-day than anything that has gone before. This was the time when there was a virtual explosion in the different types of mammals, birds, pollinating insects and flowering plants. By the end of the Tertiary, and at the beginning of the Quaternary (the Pleistocene) just less than 2 million years ago, there were more different types of living things on earth than at any other time in the history of biodiversity. At the end of the Pleistocene we come to the Holocene, our own time. It is characterized as a time of rapidly decreasing diversity, rapid climate shift, and the beginning of large-scale, highly organized activity by one dominating species – human beings. The geological timescale we have covered, together with some of the key events, is summarised in Table 1.

Table 1 Geological timescale (mya = million years ago) and some of the key events in the history of biodiversity

Era	Period	Period begins (mya)	Key events
Precambrian		4500	Origin of the five kingdoms of life, first many-celled organisms. Amount of oxygen in atmosphere increasing.
	Vendian	600	Soft-bodied organisms of the Ediacaran fauna.

Table 1 (*contd.*) Geological timescale (mya = million years ago) and some of the key events in the history of biodiversity

Era	Period	Period begins (mya)	Key events
Palaeozoic	Cambrian	544	Appearance in fossil record of all the major animal designs (phyla), including animals with hard parts – the Cambrian explosion. Jawless, heavily armoured fish. The 'wonderful life' of the Burgess Shale.
	Ordovician	505	Rapid increase in marine biodiversity. (No real land life although the ancestors of land plants are definitely present.) First fish with jaws appear – jaws and fins turn out to be a major innovation. Mass extinction (#1) at the end of this period.
	Silurian	440	First land plants and animals (arthropods).
	Devonian	410	First age of fish. First insects and amphibians (like salamanders, except some were about a metre in length) appear. Mass extinction (#2) at the end of this period.
	Carboniferous	360	Extensive forests of giant club mosses (40 m tall), tree ferns and horsetails (15 m tall) and first flying insects (like giant dragonflies the size of seagulls). Amount of oxygen in atmosphere 50% more than it is today. First reptiles (20 cm long).

	Permian	286	A time of great upheaval – extensive volcanic activity. Origin of mammals – small furry, burrowing creatures. Mass extinction (#3) of marine animals towards the end of the period – nearly everything goes extinct. Some animals, such as the trilobites, disappear for ever.
Mesozoic	Triassic	248	Ascent of the ruling reptiles (including dinosaurs). Fernlike plants dominant on land. Mass extinction (#4) at end of period.
	Jurassic	213	Ruling reptiles on land, in the air and in the sea. First birds appear.
	Cretaceous	145	First flowering plants appear, accompanied by a number of new insects (such as ants, bees and butterflies) which interacted with them, e.g. pollination. Mass extinction (#5) at end of period. Ruling reptiles and many other groups, including the ammonites, are extinct by the end of the period.
Cenozoic	Tertiary	65	The second age of fish. The age of mammals, birds, snails, insects and flowering plants.
	Quaternary	1.8	Biodiversity peaks and begins to decline – many large mammals become extinct. First humans appear. The age of humankind. Beginning of the 6th extinction.

Most scientists believe that, even allowing for extinctions, biodiversity has increased from its origins up to the present-day. Exactly why that is the case is not clear, but it must be something to do with the balance between the origin of species and their demise – extinction. These two components we shall discuss next.

beginnings of evolution – the origin of species

New species can be thought of as accidents that happen as populations of individuals adapt to different or changing environments. Although Charles Darwin entitled his book *The Origin of Species*, he did not actually explore the process of speciation very far, referring to it as 'the mystery of mysteries'. His emphasis was on showing that species change with time, not that evolutionary change necessarily results in new species. Since Darwin, considerable effort has been expended on studying speciation, but ironically more attention has been given to studying the *products* of speciation rather than the *process* itself. This said, there are a number of things we can observe about speciation in the context of biodiversity.

Many of the increases in biodiversity can be linked to times when new empty living-space appears. Someone has famously said that nature abhors a vacuum. As we've already seen, in the Cambrian the appearance of lots of new shallow-water environments was accompanied by an increase in diversity, in terms of new species and phyla. Interestingly, in the wake of the Permian there was room for diversity to flourish. However, while many, many new species appeared, there were no new phyla. The colonization and conquest of land, too, was accompanied by the genesis of many new species and groups, but still no new phyla. Clearly, despite some recent claims to the contrary, there was something quite different about the Cambrian period.

In terms of the origin of species, it would appear that the easier the access to new and/or different environments, the greater was the likelihood of new species arising – and that has happened throughout the history of life on earth. The continual movement of land masses over the planet (plate tectonics), creating, multiplying, reducing or merging continents from the Cambrian to the present day, has on numerous occasions and for a multitude of different groups been associated with the appearance of new species. The isolation of Australia from Asia more than ten million years ago resulted in a whole suite of new species – 80% of Australia's wildlife is endemic. The erection of slightly smaller-scale barriers such as mountains, hills, valleys, streams, lakes and channels has also acted in the past as a successful contraceptive, preventing different populations of the same species reproducing with each other. So two populations go their own evolutionary way and become quite different from the ancestral species. Sometimes members of a population cross or in some way overcome an existing barrier, and they form a new population which may eventually lead to a new species. This is known as the founder effect. A good example of the effect is the 800-odd fruit fly species that occur on the Hawaiian Islands. Most are restricted to a single island, but genetic analysis tells us that the closest relative of a species on one island is often a species on a nearby island, and not as you might expect another species on the same island. It looks as if there have been not just one but a minimum of forty-five founder 'events'. Climate too – or should we say, change in climate – has been implicated as a major driver of evolutionary change in past environments. When new species are produced because of separation by a physical barrier this has the technical name of allopatric (*allo* = different, *patris* = country) speciation. However, it's not just outside changes that can result in new species.

There are numerous internal changes that could also produce novelty and ultimately lead to new species. So a new species arises where there is no physical barrier – this is referred to as sympatric (*sym* = with, *patris* = country) speciation. The fact that the mutations resulting in variation are the raw material of evolution, of natural selection, means that from time to time novel features may appear in populations which could give them an advantage either in the environment they are in or in a different one nearby. The high diversity of cichlid fish in east African lakes seems to be the result of sympatric speciation. Slight specialization in different feeding behaviours of different groups, even in the same location, resulted in those groups no longer mating with others of the same species.

In terms of the timescale of speciation, Darwin put forward a theory of gradualism, which is still largely accepted down to the present-day. The accumulation of tiny changes over great periods of time leads to divergence and the origin of a new species, and over even greater periods of time, new genera and so on; but not, it would seem, new phyla, at least since the Cambrian. Stephen J. Gould, the author of *Wonderful Life*, and a co-worker, Niles Eldredge, suggested that the appearance of major differences may in fact occur over a relatively short timescale. They envisaged long periods where very little happened, punctuated by times of extremely rapid evolution. They argued that such a model (punctuated equilibrium, it was called) made better sense of some parts of the fossil record. Punctuated equilibrium is still the object of some controversy, but it has to be admitted that there are problems with the gradualist approach explaining large changes.

So we have numerous ideas of the sorts of factors that may promote or create the accidents that result in new species. We also have examples of each. Where we're still not entirely clear is which mechanisms have been most important and how common they have been throughout geological and even recent time.

end of evolution – extinction

It is alleged that one eminent professor used to start his lectures with the phrase, 'To the nearest approximation every species is extinct'. Not quite true or the study of present-day biodiversity would be very brief indeed, but it makes the point. More than 99% of all species that have existed on earth have gone extinct. Extinctions were known about even before the advent of the theory of evolution. Count Buffon (1707–78), in one of his books, talks of 'strange fossil bones ... have been found ... Everything seems to suggest that they represent vanished forms, animals that once existed and today no longer exist.' When we looked at patterns of biodiversity through time we noted numerous 'bumps'. All these drops in biodiversity, which act as bookends to each of the geological periods like the Cambrian, Ordovician and so on, are extinction events. Thus, such events define the categories we use to measure geological time. There have been five major mass extinctions.[www#22]

The end of the Ordovician may look like a little bump on Sepkoski's graph but in fact this extinction event is estimated to have wiped out 85% of all species, profoundly affecting the trilobite, cephalopod, brachiopod and echinoderm groups. The end of the Devonian saw 75% of marine species, including nearly all of the trilobites and many of the coral reefs, disappear. For some reason newly established life on land seems to have been unaffected. As we've already seen, the extinction event at the end of the Permian was the greatest so far. Many groups completely disappeared, including the last of the trilobites and the graptolites. Unlike the extinction at the end of the Devonian, both marine and land life were hit hard. Two-thirds of insect families and just over two-thirds of vertebrate families disappeared. The fourth extinction event was at the end of the Triassic period. Over a period of 15 million years three-quarters of all marine species and a good proportion of land species too went extinct. Finally, the fifth extinction event is one that we've already encountered, the Cretaceous-Tertiary boundary event. The main reason for its fame, even in popular culture, is that it saw the end of the dinosaurs and probably occurred as a result of a decent-sized meteor impact. However, both marine and terrestrial life generally were severely affected, with some groups, like the ammonites, disappearing for ever.

It would be interesting to know what brought about these extinctions. There are numerous suggestions: asteroid impacts, volcanic eruptions and other geological upheavals, deadly cosmic rays, alterations in ocean currents and dead zones (no oxygen) in the oceans, and climate change. At present, it would seem most likely that the Permian extinction was brought about by climate change and the Cretaceous extinction by an asteroid impact. The others? You can pay your money and take your choice. Perhaps the most famous (and controversial) theory was proposed by Jack Sepkoski and David Raup, both at Chicago University. They proposed that mass extinctions were not random but took place roughly every 26 million years over the past 250 million years of life's history. What may be of interest to us in later chapters is that the recovery of communities in the fossil record after these mass extinction events was in the order of 5–10 million years. The players may have changed but the number of players at this point is the same as it was immediately before the extinction event.

These catastrophic changes in biodiversity could be compared with reshuffling the cards in the middle of a card game, eliciting major reorganizations in the way life looks and how things develop from that point on. This is probably true, although it is still not generally agreed. However, while much has been written about mass extinctions, in terms of overall loss of species they haven't really been that influential. The truth is that most species that have gone extinct seem to have done so by themselves. On average, species are only present in the fossil record for 5–10 million years. This is referred to as the background extinction rate and it accounts for 96% of all extinctions. Both mass extinctions and background extinctions are almost routine events in the timeline of life. This said, the mass extinction event that could turn out to be really significant is the sixth one, the one that is taking place as you read this. And it's one that you and I are intimately involved in.

There continues to be fierce debate about the extent to which prehistoric societies impacted on biodiversity. By 10,000 years ago, and probably long before that, all of the major land masses showed signs of human occupation. There is also evidence that coinciding with the arrival of humans on a land mass was the disappearance of many of the big animals, particularly birds and mammals. This is perhaps nowhere more graphically illustrated than in the Pacific islands, where human invasion coincided with the extinction of many of the bird species, particularly if they were of the running (but obviously not fast enough) as opposed to the flying type. Both hunting and modification of the original ecosystem seem to have taken their toll of the surrounding wildlife. Human influence may have been so dramatic, even early on in our own species' history, that we may never know what a natural ecological system actually looked like. When we talk of returning things to their natural state, the truth is that in many cases we have no idea what that natural state looked like. We can only surmise, and guess. This is no new revelation. Even more than one hundred years ago, Alfred Russel Wallace wrote: 'We live in a zoologically impoverished world, from which all the hugest, and fiercest, and strangest forms have recently disappeared ... yet it is surely a marvellous fact, and one that has hardly been sufficiently dwelt upon, this sudden dying out of so many large Mammalia, not in one place only but over half the land surface of the globe.'

Certain documentary evidence for extinctions improved from the 1600s onwards, with the result that we can produce a graph of extinctions with time. But remember, almost without exception these were extinctions of generally large, easily noticeable animals and plants. Plants, birds and mammals figure large when we talk about extinction. The reason for this is simple: these are often the only groups we have good information for. It is difficult to get information for sea animals and this is borne out by the fact that only six species are known to be extinct. Similarly, we have a much better idea of what has happened on islands rather than on mainlands. After all, nearly three-quarters of all mammal extinctions have been recorded from islands. Finally, all of this information is for described species – most are not described. This is likely to be a huge underestimation. So with all these caveats and qualifications, which are so often the staple diet of science, there have been more than 1000 recorded extinctions since the 1600s, with more than half of those taking place in the last century. The number of extinctions seemed to decrease after 1994 but this decrease is not real. It is a consequence of scientists tightening up their definition of 'extinct'. From 1994 the International Union for the Conservation of Nature and Natural Resources (IUCN) defined extinct as 'when there is no reasonable doubt that the last individual has died', and extinct in the wild as 'when exhaustive surveys in known/expected habitat, at appropriate times thoughout its historical range have failed to record an individual'.

You don't have to major in biology to realize that it is extremely difficult to prove that an organism is globally extinct. Local extinctions on the other hand are much more common and we have lots of examples we could draw on: the disappearance of skate from the North Sea; the loss of sturgeon from the Caspian Sea; a reduction in leatherback turtles in California from 212,000 to 40,000 in a thirteen-year period; and a decline in worldwide bird populations to one quarter of their numbers before the advent of agriculture, to name but a few. Perhaps our best-documented example is the whales, where many species are locally extinct and many species are endangered.

I said, above, that the average life expectancy of a species in the fossil record was 5–10 million years. Well, taking information for birds and mammals, the best-documented groups we've got, the average life expectancy of a species today is, at best, less than ten thousand years. And we know that it takes 5–10 million years for biodiversity to re-establish itself after a massive extinction event. All of the books and articles that have proclaimed the coming of a great biodiversity crisis were quite simply wrong. We are in the midst of a massive biodiversity crisis right now.[www#23] What we hoped was always tomorrow has turned out to be today.

Each year the IUCN produces the 'Red Data Book', a compilation of all the information on organisms that are known to be threatened, extinct in the wild and extinct full stop.[www#24] The compilation for 2004 is given in Table 2. The global number of endangered species has increased from 10,500 in 1996 to 15,503 in 2004. One in eight birds and nearly a quarter of all mammals (23%) are still under threat since the IUCN published its first Red List of Threatened Species in 1996. In addition, a third of amphibians (32%), one in four (25%) of conifers and over half (52%) of cycads, an ancient group of plants, are also under threat. Global extinctions have increased from 766 in 2000 to 784 in 2004, with a further sixty species known only in cultivation or captivity. Some groups were highlighted as being especially at risk. For example, all twenty-one species of albatross are now threatened to some degree because of the spread of long-line fishing. Notice that while the information for birds, mammals and flowering plants on the list is pretty good, it is non-existent for some of the groups, even common and species-rich groups that we looked at two chapters ago.

In summary, extinction is a routine and integral part of what makes up today's biodiversity. It has always been around. However, the level of extinction that we are currently experiencing – extinction associated with our activities locally and globally – is on an entirely different scale. And it is to such activities that we turn in the next chapter.

Table 2 Number of threatened species (2004). Selected information from the 2004 IUCN *Red list of threatened species*

Group	2004 total	As % of described species
Vertebrates		
Mammals	1101	20
Birds	1213	12
Reptiles	304	4
Amphibians	1770	31
Fishes	800	3
Subtotal	5188	9
Invertebrates		
Insects	559	0.06
Molluscs	974	1
Crustaceans	429	1
Others	30	0.02
Subtotal	1992	0.2
Mosses	80	0.5
Ferns	140	1
Gymnosperms	305	31
Dicotyledons	7025	4
Monocotyledons	771	1
Lichens	2	0.02
Subtotal	8321	2.9
Total	15503	1

chapter five

crisis? what crisis?

If someone loves a flower of which there is only one on the millions and millions of stars, it is enough to make him happy when he looks at them for he can say to himself: 'My flower is somewhere out there ...' But if the sheep eats the flower, it is for him as if, all of a sudden, the stars went dark! And you think that is not important!

The Little Prince, Antoine de Saint Exupéry

threatening behaviour

If I were to ask what you think are the present threats to life on earth, what would you say? The way we exploit species for food, shelter or medicine? Loss of living space for living things? Human-driven climate change? Natural changes in climate or natural disasters (big asteroid!)? All but the last answer tell us about proximate causes. Exploitation and habitat destruction are threats and even if they turn out to be important threats we also have to look at what drives them in the first place, the ultimate cause – us.

Here we explore the magnitude of the ultimate cause of current extinctions – you and I, i.e. the human race. However, before

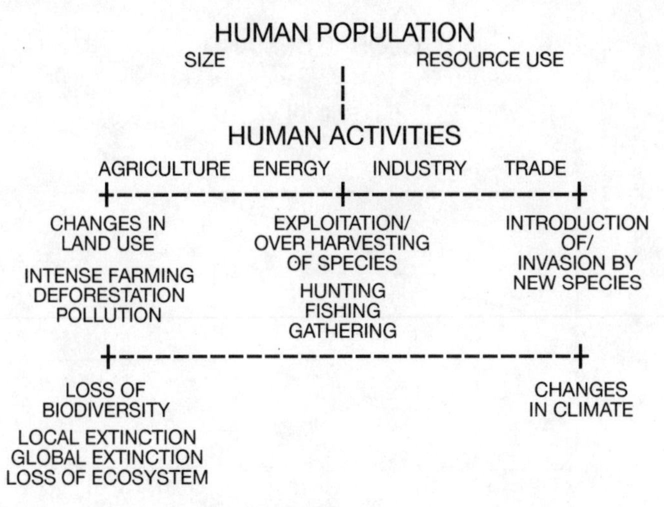

Figure 8 *How we threaten biodiversity*

this we will go through, one by one, the more proximate causes: exploitation, habitat removal and degradation, and human introduction of alien species (all laid out in Figure 8). We want to know how significant, really, is each of these threats, and the extent to which each will explain past and, if unabated, contribute to future extinctions.

living beyond our means

A recent *Living Planet Report* (2004) assessed our impact on the planet by examining what biodiversity is present and how we are using it in 149 major countries.[www#25] Figure 9 is taken from this report. It shows the ratio between the world's demand for natural resources and the world's biocapacity (resource supply, expressed as number of earths) for each year (1961–2001). The biocapacity (resource supply) of the earth is always 1 (represented by the broken horizontal line). You can see from the solid line on the graph how we have changed from using about half the planet's biocapacity in 1961 to 1.2 times the biocapacity of the earth in

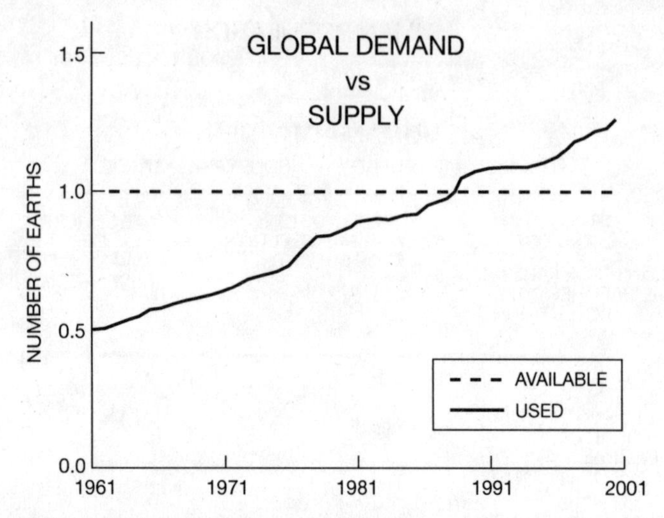

Figure 9 *World demand for natural resources*

2001. If this is true, we are currently at least 0.2 earths in deficit. Given that there is only one earth, this is not good news. We are living well beyond our means in what is referred to in the recent *Millennium Ecosystem Assessment* (2005)[www#26] as 'an unprecedented period of spending earth's natural bounty'. The assessment reported that about 60% of the ecosystem services that support life are being degraded or used unsustainably.

Presented in the middle tier of Figure 8 are the three most common known causes of animal extinctions since 1600. Remember that the animal extinctions we know about are just a small and biased subset of what is a largely unknown quantity, the total number of extinctions. And exactly how this relates to the total number of species threatened is also unclear. But the figure is still a good starting place for discussing extinctions and threats. The most common cause of known extinctions (more than one-third) is the introduction of foreign, disruptive species to new environments. This is followed closely by the destruction of habitats. In third place is direct exploitation, mainly in the form of hunting (about a quarter), with only a tiny fraction being unaccounted for. While these same three causes are seen as the major threats to current biodiversity, the extent to which they are seen as important is different. The current view is that, for all species, habitat destruction, degradation and fragmentation of natural landscapes is the greatest threat facing biodiversity. This is followed closely by direct exploitation, including hunting, and then invasive species. So this is the order in which we will deal with them.

habitat loss and degradation

Many species, and particularly tropical species, have specific food and habitat requirements. The more specific these requirements (their niche or, as the ecologist Elton put it, their 'profession') and the more restricted their habitat (where they live, their 'address'), the greater is the threat of extinction in the face of habitat loss and degradation. In the worst-case scenario there just may not be anywhere else for them to live. So the only species to do well out of habitat loss would be those that are in some way pre-fitted for the new degraded habitat, or species that are generalists and really don't mind where they are or what they eat. Nowhere on earth is untouched by human activity. Restructuring nature is a common and persistent feature of our history and activities – agriculture, logging, industry, mining, pasture, building roads, settlements and the like. Over the last thousand years we have moved more earth than all natural processes put together. Fifty years of soil loss from our planet's surface would fill the Grand Canyon. Often, what are commonly regarded as 'natural' landscapes are not. And the cost of such restructuring can be substantial. The global cost of habitat destruction has recently been estimated at US$250 billion every year, with half of an ecosystem's economic value lost when a 'natural' landscape is converted for human use. The *Millennium Ecosystem Assessment* (2005) warns that about two-thirds of the 'natural machinery' that enables life on earth – including freshwater, fisheries, air and water purification, the regulation of regional and local climate, natural hazards, and pests (we'll cover these ecosystem services in more detail in the next chapter) – is being degraded, or used unsustainably, as a result of human action.

The conversion of natural landscapes, particularly forests, for agricultural purposes has a long history. Beginning in China almost four millennia ago, it was near completed in Europe even before the time of the industrial revolution. Lowland forests began to disappear in the tropics of the New World at the same time as Portuguese and Spanish colonialism was in full swing (after 1500). Over the last 200 years considerable areas of land have been turned over to agriculture in the US, and it was in the last century that we saw rainforests, in both North and South America, targeted. More land has been converted for agriculture in the last sixty years than in the eighteenth and nineteenth centuries combined. At present, about one-quarter of the planet's land surface is cultivated and one-tenth given over to cropland. Interestingly, in some parts of the USA, on land that had reverted from farmland to second-growth forests during the great depression, the white-tailed deer population underwent a massive increase despite intense hunting pressure.

We have destroyed or degraded approximately between one-third and one-half of the world's forests. Tropical forests in particular are so important because they are home to about half of all land biodiversity. Originally there were 15 million square km of tropical rainforest. Currently there are 7.5–8 million square km, with the current rate of loss about 2% each year and growing.

And it's not just forests and woodland that are in decline. About half (more in tropical regions) of grasslands or savannahs, one-quarter of scrublands and even about one-tenth of the desert regions (hot and cold) show some sort of human disturbance. The most marked pattern of disturbance, though, is in the temperate and evergreen forests, with only about one-twentieth of their original area being classified as undisturbed. Certainly, some of the practices of intensive commercial agriculture, like the destruction of hedgerows and the elimination of weeds and insect pests, have had a massive effect on biodiversity, particularly in the more developed countries.

While habitat destruction and degradation have been investigated in some detail, the prevalence of habitat fragmentation as a threat has not received the same amount of attention. An intact forest or woodland or meadow can be sectioned into small, isolated bits by building a road through the middle, or effectively clearing strips of the habitat. Even though the amount of habitat loss may be small, the fact that you now have a number of very small, isolated 'islands' means that because of the species-area relationship (remember chapter 3 and the relationship between the number of species in an area and the size of that area?), each bit can support only a very small population. There are three problems with this. First, any negative chance events, such as disease or environmental change, will make small populations more vulnerable to extinction than they would a larger population inhabiting the same overall habitat area. Second, if any of the species present, for whatever reason, actually needs a large habitat area they are in trouble. And finally, we know that small-area habitats are more influenced by their surroundings than large-area habitats, in much the same way as it would take an ice cube one cubic metre in size much longer to melt than it would one thousand separate ice cubes made from the big one. So the way the creatures in the small bit of habitat interact with their wider environment may be completely different from the way they would interact if they lived in a big bit.

One of the practical implications of the species-area relationship is that you can also work out how many species you will lose if you reduce their habitat size by a known amount. This is related to, although greater in its scope than, the idea of fragmentation we've just considered. Thus, habitat loss has been estimated to affect nine out of ten threatened birds and plants, and four out of five threatened mammals. Specifically, we know that more than 300 bird species in Asia are threatened as a result of development, farming and logging. Information for marine creatures is understandably more difficult to acquire. Since 1980, about one-third of all mangroves have been lost, 20% of the world's coral reefs have been destroyed and another 20% badly degraded. Currently about two-thirds of the remaining coral reefs are considered as endangered, although some progress has been made in conserving them. In 1996 Brooks and Balmford reported the case of the Atlantic forests of South America, where 90% of the forest had been destroyed but no species, or at least no bird species, had yet gone extinct. Basically, because of the species-area relationship we know that the remaining 10% of forest is not enough to sustain all the species that were there. In the case of the birds, and indeed all of the species that are in the forest, many, many species are effectively extinct. They just don't know it yet. This has been referred to as 'extinction debt'.

One form of environmental degradation that has received a lot of investigation is pollution. The most serious threats come from chlorinated solvents like carbon tetrachloride, chlorinated aromatics like polychlorinated biphenyls or PCBs, components of petrol/gasoline, polynuclear aromatic hydrocarbons or PAHs, and trace metals such as zinc and cadmium. The input of chemical contaminants into the environment can potentially pose a threat, either directly or indirectly, to species and whole ecosystems. Interestingly, pollution does not seem to be a common cause of extinction globally – but it may be locally. Take a really polluted river and you find, instead of the numerous snail and insect and crustacean species that characterize clean sites, there's often just the one type of measly worm present. A relative of the earthworm and a good, abundant food for any fish that can survive the filth, these worms belong to one of the 'survive anywhere' groups. There's still life, and it may be in abundance, but it's a very poor substitute for what was there.

One of the most serious indirect effects of pollution is climate change.[www#27] By burning fossil fuels we put large amounts of the gas carbon dioxide into our atmosphere. Carbon dioxide is a greenhouse gas. It helps to trap energy from the sun's light in our atmosphere and cause it to warm up. There have been a large number of 'natural' climate changes in the past (chapter 4) but nothing like on the timescale that we're experiencing today. We've already seen that where species live (their biogeography) is greatly influenced by climate. Rapid global climate change will alter, and is altering, species' distributions. At best, climate change may merely shift these distributions; tropical species will head into more temperate regions, and temperate species to more polar regions. The main losers are likely to be the species that prefer the coolest habitats. They will find themselves, quite literally, with nowhere to go. But that's at best. It is extremely probable that many species will not be able either to adapt or to migrate as rapidly as the world is warming up. And even if migration were a possibility, many of the barriers to movement associated with our presence – roads, cities and the like – may rob species of the opportunity. Global agricultural production will be affected, but regional differences make it difficult to say what the overall outcome will be. There will be physical habitat loss as sea levels rise and many low-lying areas, including marshlands, will be flooded. Some authorities have estimated that one-quarter of all species could go extinct directly as a consequence of global climate change. The bottom line is that we really don't know, but this is not an ignorance to take comfort from. Given the seriousness of global climate change for biodiversity, and for so many aspects of human endeavour and existence, it is disappointing that many countries still do not take the threat seriously. This is perhaps best exemplified by the fact that 4% of the annual US$240 billion global energy subsidies is being allocated to renewable energy sources, while about two-thirds are spent on fossil fuels. Paying money to help destroy biodiversity or the environment is referred to as a perverse subsidy – and unfortunately there is no shortage of examples of such subsidies.

danger and opportunity?

It's become a cliché that the Chinese word for 'danger' is also the word for 'opportunity'. But taking this 'both-sides-of-the coin' approach is extremely appropriate when we probe the direct role of biodiversity 'resources' in consumption or production, thus entering the world of biodiversity as a marketable commodity. In this chapter we're dealing mainly with threats to biodiversity – and overexploitation is definitely a threat, a subsection if you like of habitat loss and degradation. But our examination of overexploitation as a threat also doubles as an introduction to the use-value, the economic value, of biodiversity, which rightly should be included in the chapter that follows this one. Use-value and overexploitation are two sides of the same coin. Which side of the coin you give attention to depends very much on who you are and what you're doing, or want to do. As Kevin Gaston and I once wrote in the first edition of our *Biodiversity* textbook, 'what to one person is "the legitimate exploitation of natural resources" can be to another "the rape of the natural world"'. So we will cover use of and threats to biodiversity together. However, this is no trivial task.

home economics

One of the 'easiest' ways into this whole area is to ask the question, 'How much biodiversity have I used today?' In my case let's start from something as mundane as my waking moment on the morning of the day that I wrote this section. I woke this morning with the sound of the bedside alarm in my ear. The alarm is made from plastic, which is derived from oil products, which themselves were once part of ancient biodiversity. The electricity powering my alarm, I think, has been generated from the burning of fossil fuels (overexploited). I pulled back my feather-filled (bird – no idea where from, overexploited or not), cotton-covered (plant – I don't know if it's overexploited or not) quilt, and lifted my head slowly off the cotton-covered (plant), synthetic-material (fossil fuels again) filled pillow. I took a deep breath of the fresh but damp air coming in my open window, air containing life-giving (and life-given, as it is produced by plants) oxygen, which fortunately enough I do not have to pay for (yet). Once up, I put the kettle on (electricity from fossil fuels), put a teabag (plant – fair trade, but I don't know if the crop – a renewable resource – is overexploited or not) into my mug, added some milk (renewable resource) and pulled down the cardboard box (trees) of Shredded Wheat breakfast cereal (plant), placed some in a bowl, added some sliced banana (plant) and raisins (plant – this is getting tiresome) and reached for the milk carton (cardboard – tree, covered with plastic – fossil fuel), wishing someone would make me kippers (overexploited) instead. All of this even before the kids get up or I've gone for a pee.

This catalogue of domestic trivia makes an important point. If it is difficult, perhaps near on impossible, to get to grips with how much biodiversity one middle-aged academic uses between getting up in the morning and eating his breakfast, how much more difficult will it be to begin to comprehend the scale of our direct use of biodiversity globally? It is simply breathtaking and bewildering at the same time. Direct use is complex and multifaceted and often it is difficult, if not impossible, to tease apart those different facets. And at times we simply do not have the information that allows us to know where the ever-shifting line between exploitation and over-exploitation lies.

food, glorious food

Biodiversity is the basis of all food industries, both directly and indirectly. It is the basis of all the service industries too. The production and consumption of food is essential for human existence and so, not surprisingly, is also very big business.

Much has been made in some textbooks on the (bio)diversity of foodstuffs. For example, someone has reported that up to 500 food-plant species have been recorded in home gardens of one village in Java. And about fifty different types of acacia seed have been used as food by Australian aboriginals. In reality, though, of the 12,500 plant species considered edible, only 200 have been domesticated and of those three-quarters of our total food supply comes from just a dozen different kinds. And global agriculture is responsible for more than nine-tenths of our protein intake. There may not be many kinds of plant contributing to our food needs but the scale of the exploitation is enormous, at 2.7 billion tonnes produced every year.

Fungi are used extensively as foodstuffs, with many hundreds of edible varieties worth an estimated US$700 million in the US alone. In the UK even just one fungus species, which is sold as Quorn (a low-cost protein source and often a meat substitute), is worth £25 million.

As with plants, the use of animals as food is restricted to a very small proportion of the total of animals that could potentially be used. Much of our use is culturally determined. Vertebrates – that is, fish, amphibians, reptiles, birds and mammals – are high up in the list of animals used as food, as are substances derived from them, e.g. cows' and goats' milk and cheese. To a lesser extent, and depending on where you live, there are also molluscs (such as clams, mussels, snails, squid and octopus), crustaceans (such as lobsters, shrimps, prawns and crabs), insects (beetles, moths and the produce of bees – honey) and even echinoderms, some sea cucumbers and a few sea urchin species. While average global meat production is high, it is still only one-tenth of edible plant production at just over 200 million tonnes every year. Annual extraction from marine and freshwater environments is in the region of 100 million tonnes every year, worth an estimated US$60–100 million.

The other side of the coin is that many of the uses of biodiversity as food are unsustainable, at least in their present form. The history of hunting is such that often species have been pursued right up to the last viable population, or even the last individual. This is very much the story of the whaling industry. Not surprising then, as we've already seen, that hunting is the third most common reason for recorded extinctions (although not just for food but also for fashion). One of the best examples of unsustainable practice is the hunting for bush meat in some tropical forests. The Amazon basin produces 1.8 million tonnes of mammal meat annually. Of this only 150,000 tonnes are hunted, which means that you only need around a 15% increase in production rate to make up for the bush meat taken as food. Contrast this with the Congo Basin, where the annual production rate is just slightly greater than in the Amazon at 2.1 million tonnes of mammal meat annually. Here 4.9 million tonnes are hunted every year. You don't need to be a mathematics graduate to work out that this is completely unsustainable.

Another high-profile example of how our use of biodiversity for food can be unsustainable is commercial fishing. The middle of the last century saw a massive increase in global fishing effort. It is claimed that the first major well-documented fisheries collapse was the Peruvian anchoveta fishing industry in the early 1970s. But there are a large number of earlier cases which often had profound effects locally. I even remember as a child in Glasgow in the 1960s my mum chatting with others about the disappearance of the herring from the shops and from the local seas. Currently, average global fisheries production is in excess of 80 million tonnes every year. Of the 200 most important fish stocks (77% of the total catch), 35% are thought to be overexploited, with, in some cases, catch reduced by two orders of magnitude from its level before the industrialization of fishing. The ocean's large predators, the sharks, the tuna, and the swordfish, are now an order or magnitude less abundant than they were.

However, apart from their obvious impact on the commercial species of interest, fishing activities have a much wider impact on biodiversity. Massive reduction in one species invariably has knock-on effects for many other species. For example, overexploitation of herring and capelin in the north-eastern Atlantic led to the collapse of many large fish, sea-bird and sea-mammal populations. Moreover, the methods used to catch fish are rarely selective. Often many tens or even hundreds of different species, both fish and invertebrates, are caught 'accidentally'. I took some photographs of all the animals present in a bottom trawl that was being carried out to catch scampi (Norwegian lobster, *Nephrops*). Scampi are bright red, so relatively easy to see. But in a half-metre-high pile consisting of many, many hundreds of sea urchins, starfish, swimming crabs, dead man's fingers, clams and brittle stars (indeed, almost every animal phylum is represented) there were only nine scampi. Other fishing methods for other species can result in the capture and subsequent death of even larger creatures such as turtles and dolphins. Overall, such by-catch can be substantial – in the order of a quarter of a million tonnes annually. This is to say nothing of the habitat destruction caused by such methods as bottom trawling, and the environmental effect of lost nets and lines and the like. There are some areas in the Irish Sea, the North Sea and the English Channel where the scouring and subsequent rescouring of the sea bed is one of the persistent features of the landscape.

While many of our current food production and consumption practices are unsustainable, it is worth keeping in mind that global food demand is expected to more than double by 2050.

fuel for the wheels of industry

A tremendously wide range of industrial materials is provided by biodiversity. They range from building materials like wood or grasses through to glues, dyes and even rubber and perfume. The scope of this facet of exploitation is so vast it is almost impossible to comprehend. To make the point, we'll focus on one aspect where quite a lot of information is readily available. Three billion, eight million cubic metres of wood are harvested every year. But wood is not only used as an industrial material. Two billion people – one-third of all humankind, and its poorest members at that – are reliant on fuel woods and these alone as a source of energy for cooking and keeping warm. There is currently an estimated 2.5 billion cubic metres of fuel wood or equivalent. Even allowing for many hundreds of schemes to increase the amount of fuel wood available (tree planting, for example) the demand by 2010 is estimated to be between 2.4 and 4.3 billion cubic metres. This is only one aspect of exploitation and over-exploitation of industrial materials, but the figures speak for themselves.

medicine *sans frontiers*

Biodiversity provides many of the world's medicines and drugs and has done so for millennia. One of the earliest must be the opium poppy, as one of its chemical constituents is the powerful painkiller morphine. Since four out of five people in the world do not have access to modern medicine, natural products from medicinal plants play a key role in the health of most of the world's population. About 20,000 plants are used in traditional medicines. Even many of the commercially available medicines have their origins in biodiversity. Aspirin is derived from acetylsalicylic acid present in the meadowsweet plant. The antibiotic Penicillin comes from the fungus *Penicillium.* Of 520 new commercial drugs introduced between 1983 and 1994, about two-fifths were made from natural products or their derivatives. And of the twenty best-selling drugs, about half were made from natural products or their derivatives, with a monetary worth exceeding US$16 billion. The best seller ever (greater than US$1 billion a year) contains taxol, which is used to treat breast and ovarian cancers. Taxol was isolated from the Pacific yew tree, a species which used to be discarded as worthless during logging operations. The hit rate for producing new drugs is better in nature than it is in the laboratory. One in 125 plant species produces a major drug, compared with one in 10,000 specially made chemicals. Even some of the more obscure animal groups that barely got a mention in chapter 2 can turn out to be important in producing new drugs. For example, drugs for the treatment of cancer have been isolated from sponges (phylum Porifera), sea mats (phylum Bryozoa) and sea squirts (phylum Chordata). It has been calculated that 50% of all 'substances' isolated from marine animals (e.g. shark's liver) and plants since the early 1970s have anti-cancer properties.

The other side of the coin is that there is the possibility of over-harvested medicinal plants or animals becoming endangered or even going extinct. Today, even bits of currently very rare animals such as turtles, tigers and rhinos are in great demand for some traditional medicines, often irrespective of whether or not they possess any medicinal qualities. It is extremely difficult to estimate the global threat to biodiversity of collecting animals and plants for medicinal use, as most extinctions will probably occur on a greater scale through habitat destruction and degradation.

ecotourism

Travelling to locations to see and experience wildlife is known as ecotourism.[www#28] Not only does it use biodiversity as a resource but its existence is based on the concept of biodiversity. Ecotourism is a relatively modern phenomenon and is also big business. In 1998 nine million whale watchers worldwide spent an estimated US$1 billion on the pursuit. Viewing elephants in Kenya was worth an estimated US$25 million. Even a quarter of a century ago, in one year there were upwards of 250 million international ecotourists (still only one tenth of the total if we include those who travel within their own country), contributing US$233 billion to national incomes. To get an idea of the scale of national ecotourism recreational use of public forests in the US is worth US$16–24 per visitor per day, while in the UK the value has been estimated at US$79.5 million per year. Also, you don't have to travel very far to qualify as an ecotourist: the *International Zoo Yearbook* (1990) reported that more than one-tenth of the world's population had visited a zoo.

While it is true that ecotourism, by raising awareness of the existence of, and threats to, biodiversity is potentially a very positive development, there could be some significant downsides. While responsible ecotourism could deliver real environmental and social benefits it could also be taken merely as a new marketing opportunity – what has been christened 'green greed'. Even with the best motives in the world ecotourism may have a negative impact on the wildlife or environment at its centre by increasing pollution and transportation to and from the destination and degrading the environment as facilities need to be built to house and entertain ecotourists. Sheer numbers may well overwhelm the destination and exceed the carrying capacity of the area. Many reading this chapter will be keen to know, given everything said above, whether ecotourism helps or harms the environment. The answer is undoubtedly both. What is not clear is which outcome is most common.

controlling the natural world

Use of natural enemies to control problem species is a common feature of biological control. Leafy spurge is a weed introduced into the US in 1827. It displaces native plants and restricts cattle grazing. Biological control of this plant began in the 1960s and has entailed the introduction of fifteen foreign (non-indigenous) insect species. The most effective have been the flea beetles, which destroy 80–90% of the plant in an area. The economic benefits have been huge but have not been properly assessed.

This is not the case for the control of the red waterfern, where scientists have made detailed study of the economics of control. This waterfern is native to South America but it invaded waterways in South Africa, causing economic loss to water users, mainly farmers but also those using waterways for recreation. The frond-feeding weevil was released as a biological control measure in 1997 and within three years the waterfern that it fed upon was largely under control. The economic cost of the 'damage' caused by the invasion was US$589 per hectare per year. The cost of biological control was US$278 per hectare per year, with an initial one-off investment cost of US$7700. So in 2000 the benefit:cost ratio was estimated at 2.5:1, but by 2005 had increased to 13:1 because the treatment had been so successful. Some other success stories are the control of cassava mealybug in Africa by an introduced wasp species, with annual savings in excess of US$250 million, and control of the banana skipper butterfly in Papua New Guinea, with annual savings of around US$148 million.

In summary, when it is successful, the economic gains of biological control can be huge, but when it goes wrong, it really does go wrong. This is intimately linked with the next major threat we need to discuss – the introduction of species to new areas.

introduced species

When I first visited Wembury beach to study intertidal prawns in the summer of 1986 I was asked to look out for an invasive seaweed that was reported to be working its way around our coasts. Sure enough, some pools had this *Sargassum* weed that had originally come from Japan. When, fifteen years later, I came to live in Wembury I was shocked to see how many of the intertidal pools were dominated by *Sargassum*, reducing the light available for other seaweeds and totally altering the character of the pools. After habitat loss and direct exploitation, the biggest threat to local and global biodiversity is believed to be invasion by non-native species. The invaders, as we've already seen, are sometimes introduced deliberately for the purposes of biological control. Others are introduced, both intentionally and unintentionally, mainly through trade and tourism. Approximately 400,000 species have been introduced; only one in ten of these has become successfully established. Of those established species, only a further one in ten goes on to become a pest. That's not a great success rate for the making of a pest, but, still, a very small number of pests can play havoc with their new environment and its inhabitants.

It is important to realize that invasions have always happened. What is critical now is the vast increase in their number as a result of increased human activity. For example, a colleague of mine from Sheffield University studied the types of seeds present in mud on the wheels of lorries coming into Plymouth from France and Spain, and there is work which shows that the number of invaders in nature reserves is directly related to the number of human visitors coming into the reserve.

Invasive plant species already cover 400,000 square kilometres of the US. They continue to spread at a rate of about 12,000 square kilometres a year. Half of all the threatened species in the US are thought to be at risk because of invaders. The 50,000 invaders currently in the US are thought to cost about US$137 billion, both in economic damage and in attempts to control them. The Galapagos Islands have almost as many introduced species as native ones. The Baltic Sea now has an additional 100 new invasive species, a third of them native to the American Great Lakes. This said, a third of the 170 aliens in the Great Lakes are originally from the Baltic.

Pests can play a major role in biodiversity decline. These invaders either eat the threatened species (predators) or live off them (parasites). The great lakes of Africa – Malawi, Tanganyika and Victoria – were famous for the large number of endemic species of cichlid fishes they supported. However, someone had the bright idea of introducing a really big fish-eating predator, the Nile perch, into these waters because it would improve the fishing. The Nile perch has become established in Lake Victoria and is quite simply eating to extinction many of the cichlids.

I accept that it is easy to cast a large predator in the role of 'baddy', but even cute, fluffy creatures can ravage whole ecosystems in a less-than-cute-and-cuddly way. Take the European rabbit, for example. It originally occurred on the Iberian peninsula of south-west Europe. However, the ancient Romans introduced it to Italy, not because it was cuddly but because it was tasty. Rabbits – at least, those that weren't eaten – did well in Italy and in other European countries. They came to England in 1066 along with the Norman conquest and thereafter did a bit of island-hopping during the Age of Exploration.

The classic 'thing-not-to-do' was done by Thomas Austin in 1859. He was an English landowner who had emigrated to Australia. I remember missing the BBC news and brown HP sauce when we lived in Canada. Austin specifically missed blowing bunnies apart for fun: so he imported twenty-four individual rabbits to Victoria. Today, rabbits are one of Australia's greatest environmental problems. They are the main cause of habitat destruction and degradation. They devour crops. Sixteen rabbits eat as much as one sheep does. In a 'good' year there can be a billion rabbits. They alter whole ecosystems just by how they feed. They threaten elements of the native fauna with extinction – in particular, marsupials such as the Greater Bilby (now endangered) and the Burrowing Bettong (now extinct on the mainland). They directly compete for food and habitat, ousting many, like the Rufous Hare-wallaby, from their burrows. Rabbits cost Australians US$600 million every year in control and loss of production.

A good example of where we have a successful invasion but the invader turns out to be quite benign is the terrestrial 'shrimps' originating from the southern hemisphere. Commonly referred to as 'landhoppers', these shrimps have successfully invaded leaf-litter habitats both in Europe and in the US. One species that I have investigated quite extensively is an Australian, *Arcitalitrus*, which now is very abundant in areas of south-west England and a couple of islands off the west coast of Scotland. Landhoppers are present in woodland adjacent to Wembury beach, and they thrive in our back garden at home. This species can occur in densities of many tens of thousands per square metre but they do not seem to have any negative effect on the native fauna, except for being a nuisance by invading our house when it rains a lot (I removed one from my bath the other day). We thought maybe it competed with woodlice, but there was no evidence that it did. We compared the identity and abundance of animals that lived in areas where landhoppers were common and areas where they were absent: there does not seem to be any difference. Having studied them since 1986, we have uncovered no negative effects of landhoppers on the natives. And that seems to be the case for nine out of ten successful invaders.

the domino effect – extinction cascades

It is not uncommon to find that the extinction of one species leads to the extinction of one or more different species. This is called an extinction cascade. For example, a species that is right on the brink of extinction is the *Calvaria* tree from the island of Mauritius. Passing through the gut of a flightless bird, the Dodo, was the key to the germination of the seeds of this plant. The Dodo was hunted to extinction in the late seventeenth century. Since then *Calvaria* seeds can still be found but they cannot germinate. The species is now hanging on, just, in the form of a few very old trees.

A recent study by Lian Pin Koh and his colleagues (2004: *Science*, 305,1632) modelled co-extinction (the loss of a species upon the loss of another) for a set of interrelated species, including wasps, parasites and their hosts, butterflies and the plants that fed their caterpillars, and ants. Based on what they found, they estimated that for every one species on the endangered list, there are 6300 related species that are 'co-endangered'. If they're right, we will have to revise the numbers of endangered species that we discussed in chapter 2.

Not all extinction cascades are global. There are also 'local' examples where the extinction of a species in one area has dramatic knock-on effects for all the other species in that area. Linked with the threat to biodiversity by invaders that we've just dealt with, above, there is the story of the mysid shrimp (sometimes called opossum shrimp) that was introduced into Flathead Lake, Montana, US. The shrimp was supposed to be an extra food source for the economically important salmon in the lake. Unfortunately, the shrimp ate very large numbers of the native microcrustaceans. The microcrustaceans were the mainstay of the salmon, so the salmon numbers collapsed to such an extent that populations of bald eagles, brown bears and human residents that relied on the salmon were all adversely affected.

some light relief – complete elimination of biodiversity by extraterrestrial means

'You can not help but get a big flash when objects meet at 23,000 miles per hour,' said Dr Pete Schultz when, at 1.52 am on 4 July 2005, the NASA spaceship *Deep Impact* slammed into Comet Tempel 1. 'The heat produced by impact was at least several thousand degrees Kelvin and at that extreme temperature just about any material begins to glow.' The *Deep Impact* mission was all about investigating the origin of the solar system, studying the make-up of the comet by throwing things at it. However, it also illustrated very graphically how something the size of a refrigerator hitting an object floating in space can cause an awful lot of damage.

Consider, then, that there are 1100 objects, asteroids or comets, greater than 1 km in diameter, and more than a million greater than 40 m in diameter (modest office block size) which regularly come close to earth, and hold the possibility of striking our planet. These Near Earth Objects (NEOs), if they cross our path and get through our atmosphere, could cause widespread environmental damage and biodiversity loss on a local (if a 40 m–1 km diameter NEO) or even global (if a 2–25 km diameter NEO) scale. None of the currently recognized NEOs (which can be found updated daily on a NASA website) is thought to be a threat, but how do you predict the behaviour of a presently unknown object? On average, a 'dangerous' NEO (2 km in diameter, 1 million megatonnes of energy) does collide with the earth once or twice every million years. Currently you, and other bits of biodiversity, have a one or less chance in a hundred million of being killed by an asteroid collision. But don't let it keep you awake at night. Despite an estimated 100 people worldwide, and the NASA-backed Spaceguard Programme, looking out for NEOs, your first indication of an impending impact is likely to be a spectacular short-lived (for you, at least) flash and shaking, not at all like the Hollywood movies *Deep Impact* or *Meteor*.

In 1980 the physicist Luis Alvarez and his son Walter, a geologist, proposed that an asteroid 6–15 km in diameter (100 million megatonnes energy) collided with the earth about 65 million years ago (coinciding with the extinction at the end of the Cretaceous period). This resulted in the Chicxulub crater on Mexico's Yucatán Peninsula. The impact would have penetrated the planet's outer covering, much as we saw with the *Deep Impact* experiment, scattering dust and debris into the atmosphere, and causing all sorts of natural disasters. Again, as in the *Deep Impact* experiment, the heat produced would have incinerated all life in its path. So death by asteroid is seen as the most likely cause of the extinction at the Cretaceous/Tertiary boundary.

In an article in the magazine *Science* (May 1991) evidence was presented for the rapid demise of most life forms (over a period of 10,000 years rather than the 10 million that used to be proposed) in a mass extinction event at the end of the Permian period. It has been suggested that the late Permian, late Cretaceous and many other mass extinctions could have had an extraterrestrial cause, with the slate of life almost being wiped clean every 100 million years or so. Palaeontologist David Raup, of the University of Chicago, has suggested that roughly 60% of all species extinctions may have been caused by impacts.

So the possibility of extraterrestrial objects hitting the earth has been a remote but serious threat in the distant past, and continues to be so in the present day. It is entirely possible that much of the history of biodiversity has been shaped to some extent by such collisions. The extent to which biodiversity will be 'modified' by such impacts in the future is unknowable. And this is the one threat to biodiversity where, if the light around us gets suddenly brighter, we can rest assured that it's genuinely not our fault.

once upon a time there were two people ... now look how many there are

For most of the three million plus years of our existence, humans have lived short and often precarious lives, as *ad hoc* exploiters of biodiversity – what we call hunters and gatherers. However, as we've already seen, even a prehistoric world population of around ten million people could have a major negative impact on biodiversity. A more organized approach to exploiting biodiversity began to emerge 6000 years before the present, at first in the Middle East – this was agriculture. This new lifestyle resulted not just in different kinds of, but larger-sized, communities. By the first century AD the world population was estimated to be 300 million, growing steadily to 760 million by 1750. The advent of the industrial revolution resulted in an increase in living standards, for some at least, and coincided with a respite from the more severe famines and epidemics that had until then plagued communities. Thus, by 1800 the world population had reached the one billion mark, two-thirds of which resided in Asia and a further one-fifth of which lived in Europe. Although it took a few million years to reach the first billion, the second billion took only 150 years. Between 1960, the year that I was born, and when I turned fifteen another billion were added! This increase was to some extent fuelled by global attempts to reduce infant mortality (e.g. introduction of immunization programmes against diphtheria, cholera and other life-threatening diseases) and enhanced global food production as a result of the use of fertilizer and more efficient farming practices and the introduction of new disease-resistant crops (e.g. new strains of rice). The twentieth century

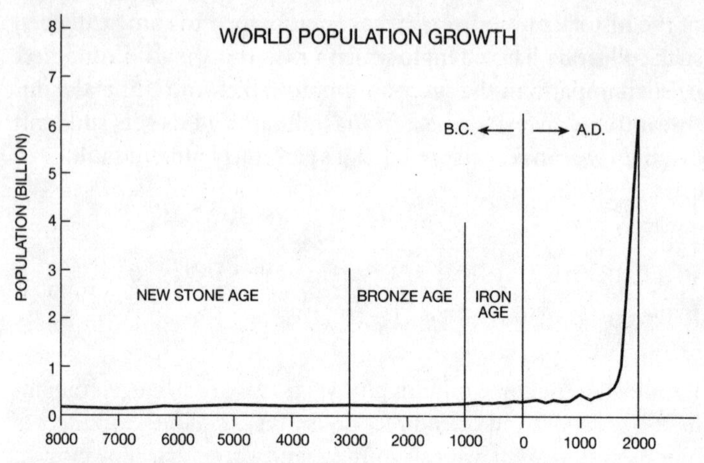

Figure 10 *World population growth*

started with 1.6 billion souls on the planet and finished with 6.1 billion. The human population has multiplied more rapidly than before,[www#29] and this exponential growth seems set to continue (although growth rates are showing some reduction owing to declining birth rates), with a predicted 9 billion by 2050 (Figure 10). Most of this growth is forecast to occur in what we term the less developed countries – in Africa, Asia and Latin America. Africa, in particular, will see huge increases in population, although many of the current calculations do not always take into account the impact of the HIV/AIDS tragedy.

And right now, three children are born every second. Each day a quarter of a million children are born. We saw earlier that ten million of our ancestors had a profound and irreversible effect on biodiversity; consider now that the total number of all those ancestors represents a mere forty days' worth of births at current rates.

Many regions of the globe cannot produce enough food to support their populations. One place where this is glaringly obvious is equatorial Africa. The whole area is quickly being transformed into a desert. The fast growing population is rapidly clearing the forest for agriculture and for fuel wood. About a century ago, half of Ethiopia was forest; today, only about 4% is forested. Sub-Saharan Africa has the highest birth rate and shows the greatest rate of population increase of anywhere in the world. The population will double by 2016, but food production will not. Such rapid population growth cannot but lead to greatly increased rates of habitat loss and fragmentation. However, we would also do well to remember that at the moment the most pronounced pollution and energy-hungry 'hotspot' countries are in Europe and Asia.

While human population size is considered the foremost driver of biodiversity loss, little attention has been paid to how this is related to the number of households that exist. More households means greater consumption of wood for fuel, greater habitat alteration for home building, and increasing greenhouse gas emissions. Lui and his colleagues (2000: *Nature*, 403, 853) found that household numbers had increased globally (1985–2000), and particularly in those countries containing biodiversity hotspots. Even in cases where population size declined, the number of households still increased substantially. These investigators proposed that a reduction in average household size alone will add a projected 233 million additional households to hotspot countries (2000–2015). Such proliferation in household numbers translates into urban sprawl, greater per capita resource consumption and so is a further drain on biodiversity

Finally, it is not just total numbers, but the distribution of population and what that population does that matters. Migration into environmentally sensitive areas has a big impact, while concentration of human activities in towns and cities focuses pollution on to small areas. Over the period 1980–97 global increases in consumption were nine times greater than the population increase. The scale of this consumption is difficult to comprehend. Figures produced by the International Energy Agency reveal that in the ten years ending in 1997, energy consumption in North America rose by 16%, in South America by 35%, and in the Middle East and North Africa by 58%. Humans use a quarter of total rainfall and about one-half of all available freshwater running off the land, mainly for agriculture. Water withdrawals from lakes and rivers has doubled in the last forty years. The flow from many rivers has been reduced dramatically with, at times, water in the Yellow River in China and the Colorado River in North America (among others) never reaching the ocean. About a-billion-and-a half people live beside rivers that experience serious water shortages.

In terms of energy consumption, this has increased by three orders of magnitude from the time of the agricultural revolution up to the present day. Present-day commercial energy production is the equivalent of 163 million barrels of oil. Meanwhile, in the ten years to 1999, vehicle miles travelled by car in the US and the European Union rose by nearly 80%. Globally, air traffic doubled.

It is inconceivable that a venture on this scale would not have major detrimental impacts on biodiversity.

it's the poor who do the suffering

As well as the numbers of human beings increasing exponentially, the economic fate of these new additions is also changing both rapidly and dramatically, with major implications for biodiversity. In 1960, when I was born, one-fifth of the world's population had two-thirds of the world's money. By the time I left the UK to take up my post-doctoral fellowship in Canada in 1989, that two-thirds had increased to over four-fifths. During that same period the poorest fifth of the world's population saw a reduction in their income from just over 2% to around 1% of global income. It was estimated that in 2002 the richest 1% of people in the world earned the same amount as the poorest 57%.

About 1.3 billion people live on less than US$1 a day. Contrast this with the money locked up in the world's tax havens – US$11.5 trillion. That's more than enough to pay for UN Millennium Development goals (chapter 7) and to eradicate poverty in Africa twelve times over. Figures published by the United Nations in 2002 showed that the net income of the three richest families in the world (the Gates family of recent *Live8* fame, the Sultan of Brunei and the Walton family) is the same as the GDP of the world's poorest forty-three countries combined.

The US, with one-twentieth of the world's population, consumes about a quarter (or more?) of the world's resources and generates a quarter of the world's waste. Each US citizen consumes the same amount of energy as 35 Indian citizens, 150 Bangladeshi citizens and a largish village of 500 Ethiopian citizens.

By and large, countries of the southern hemisphere tend to be poorer than those in the northern hemisphere. There is a latitudinal gradient in poverty which mirrors the latitudinal gradient in species richness that we saw two chapters ago. The greatest poverty coincides with the greatest species richness. But irrespective of where you are, the extent to which a country's earnings are skewed towards a small number of wealthy individuals (income equality) has increased over the years. In 1979, the richest 20% of US citizens earned nine times more than the poorest 20%. Eighteen years later they were earning fifteen times more.

The United Nations has estimated that more than 100 million people living in wealthy countries, including more than 15% of Americans, live in poverty. Brazil, too, shows perhaps the greatest gulf, with nearly one-half of the nation's income going to one in ten of the earners. In every way you look at it, the rich get richer while the poor get poorer.

While a rapidly growing population cannot but impact on biodiversity, it is equally true that it is what those people actually *do* that has the greatest influence. Inequalities in wealth distribution seem to result in a huge scale of exploitation and pollution by the rich, and perpetuation of poverty as the poor are forced to degrade their natural environment in order merely to subsist. Damage to ecosystems impacts most directly on the poor. It is the poor who suffer the greatest effects of polluted environments. It is the poor who suffer the loss of productive lands. It is the poor who suffer the loss of traditional sources of food, fodder, fuel and fibre when forests are destroyed. The poor suffer a greater vulnerability to local environmental degradation, which money to some extent alleviates.

Recent advances in economics and in technology only seem to widen the economic gaps both within and between countries. The global economy is forecast to be two to three times its current size by 2025. If such growth is to be realized, and we are genuinely to attempt to 'Make Poverty History', we need access to the equivalent of three 'current earths' with associated biodiversity. All other threats to economic growth, world peace and quality of life issues seem small fry compared with this.

So, the effect of declining biodiversity impacts on the world's poor to a greater extent than the rich. The conservation and management of biodiversity is therefore, arguably, more critical for the poor than for the rich because, quite literally, biodiversity use is more 'immediate' in their lives and livelihoods – it can be a matter of life and death even in the relatively short term. Faced with the complicated mixture of threats and problems we've looked at in this chapter, with all their political, sociological, moral and biological implications, the question is exactly what *do* we and what *should* we value about biodiversity? This is the subject of the next chapter.

chapter six

no such thing as a free lunch

If you were to mention to grown-ups: 'I've seen a beautiful house built with pink bricks, with geraniums on the windowsills and doves on the roof' ... they would not be able to imagine such a house. You would have to say to them: 'I saw a house worth a hundred thousand pounds.' Then they would exclaim: 'Oh! How lovely.'

The Little Prince, Antoine de Saint Exupéry

'... and for everything else there's mastercard'

When I got engaged in 1979 to Fiona, now my wife, I bought her a beautiful (but modest – I was still a student) gold ring. In return she gave me a pure white Antoria acoustic guitar. The guitar is a bit battered and worn now. Its present market value is comical. However, indirectly through the years it has more than paid for itself through gigs and functions, and it has brought (to me, at least) inestimable pleasure. My guitar is worth very little and at the same time, it is priceless. Given how incredibly difficult it is to put a value on something as simple as a guitar, where do we begin when we want to ask the question, 'What is the value of all living things?' This is a particularly pressing question when we consider the enormity of the threats. So how do or how should we value biodiversity?

As with the guitar, it is nearly always impossible to disentangle satisfactorily all of the different types of value we may attach to biodiversity. Take the temperate rainforest on the west coast of Vancouver Island, Canada, where my family (and guitar!) spent some time in the late 1980s. Certainly, real hard cash was earned from logging and associated activities, as well as fishing (big salmon!), hunting, recreation (great walking and trekking, with the famous West Coast Trail which ended at our door), tourism and ecotourism (lots of whale watching). But indirectly, and a little more difficult to put a monetary value on, the existence and well-being of the rainforest was essential to protect the many watersheds and keep the soil in place and 'working', with associated implications for water cycling and management, the production of oxygen, the 'mopping up' of carbon dioxide (a greenhouse gas) and many other free (to us) but essential biological 'services' – often termed ecosystem services. Even more difficult to quantify, and inextricably linked with all of the above, is the 'value' derived from the rainforest in terms of the cultures (indigenous and non-indigenous) it has helped to shape, the art, music and literary work, even the religious beliefs it has inspired. And how do you put a price on one of the most incredible experiences I have ever had – from the top of a mountain in the forest, watching the sunset over the trees, and through those trees trace the hundreds of broken islands out to the west?

In some ways it seems wrong to want to tease out all of these different interrelated values. And yet against the backdrop of unprecedented losses of biodiversity in recent times, and the continuing pressure on biodiversity for profit, for livelihoods, for mere existence in some cases, we do need to try to evaluate all of these different values, however unsatisfactory that might be. The order in which we tackle the various values of biodiversity could be taken to indicate their value to the author. It does not. We'll use a generally agreed scheme devised in the mid-1990s for dissecting the different values of biodiversity.

costing a small planet

The value of biodiversity is divisible into a number of different categories (Figure 11). None of these is watertight, and in reality there is much blurring, and not just at the edges either. The total economic value of biodiversity can be divided into two, largely self-explanatory, categories – use value and non-use value. Use value can be further subdivided into utilitarian value (useful now), option value (possibly useful soon) and bequest value (useful to future generations). Utilitarian value can be further subdivided. The value of marketable commodities (production and consumption subject to direct trading) is perhaps the easiest to handle. We've already covered this in the previous chapter when considering threats to biodiversity. The value of non-market commodities – aspects of biodiversity which are not directly subject to market trading, but could still in theory be given an estimated market value – is what we will start with here. Starting in the section that follows with an attempt to put a price tag on non-market commodities (what the earth gives us for free), we then move on to look briefly at option and bequest values. To finish the chapter we'll consider aspects of two sets of values that are often so intertwined we may never be able to separate them. Existence value, 'satisfaction' that a resource is

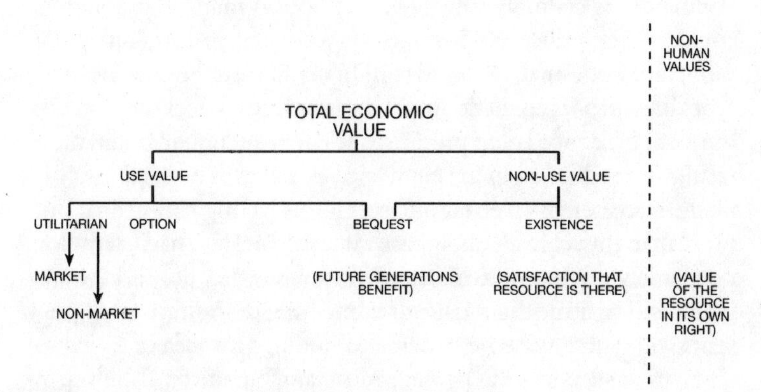

Figure 11 *The value of biodiversity*

there, is one of the key non-use categories, and there is also intrinsic value, the value of a source in its own right independent of human values. We'll look at what different people and faiths have believed about biodiversity through the ages as a way of attempting to understand existence and intrinsic value, as well as throwing some light on our overall approach to biodiversity.

use now, pay when?

There are many things that biodiversity *does*, things that are essential for maintaining a planet that is capable of sustaining life – and in the process keeping us alive. Making and maintaining our atmosphere, recycling nutrients, controlling species that could be pests, introducing energy from the sun into our world through photosynthesis, pollinating plants – these are a few of the services from nature that we perceive as being 'free' and that come under the umbrella-heading of non-market or indirect use value. Now, while it is true that many of nature's 'services' are not subject to direct buying and selling and so do not (yet) have a monetary value, we do not exactly get them for free.

Take, for example, the indirect value of pollination, the reproduction method of many seed plants. This involves the transfer of genetic material between plants by insects and, in some cases, birds. Pollination is certainly valuable as without it many plants simply could not reproduce. No harvests, no food. But you couldn't put a monetary value on it. Or could you? In many areas the numbers and types of wild bees are in decline. This is having a knock-on effect in the form of reduced crop production. The reduction in pollination results in a reduction in the plant species that depend on these pollinators, so there is an economic cost to *not* having pollinators. But more than this, consider how costly it would be if we had to employ people to work in our fields manually exchanging pollen between plants. All of a sudden what was something of indirect value presents you with a fairly hefty demand notice. This idea of trying to cost nature's services was perhaps most tangible and hard-hitting in what has become a landmark scientific paper.

Robert Costanza and twelve colleagues, writing in the magazine *Nature*, suggested that it might be useful to attempt to put a monetary cost on nature's services, the ecological systems critical to maintaining life on earth. Taken together, these services are the basis for all human welfare. As such they must, in some way, be part of the total economic value of our planet. So Costanza and colleagues took published information, and made a few of their own calculations. They estimated the 1997 economic value of seventeen ecosystem services. Scaled up for the whole earth, the total value they calculated was around US$33 trillion per year (Table 3). Not only was this, they claimed, a conservative estimate but nearly all of that 'value' was outside the current market. So, assuming we could actually buy the things that the earth does for us for free, could we afford them? In the Christian New Testament Jesus tells a parable of a man who finds a most beautiful product of biodiversity, a perfect pearl. The man then goes and sells all he has to get that pearl. Our 'all we have' approximates to about US$18 trillion per year – the gross national product of the earth – and that's only just more than half what we need for our pearl – what we need to survive and what biodiversity gives us for 'nothing'.

Table 3 Global value of ecosystem services as estimated by Costanza and his colleagues (1997). Notice that, somewhat counter-intuitively perhaps, cropland is not worth very much compared to the others, whether you compare it per hectare or take it altogether. Wetlands are by far the most valuable ecosystem per hectare, and coastal zones in terms of total value.

Ecosystem	**Area covered** (hectares)	**Relative value** (per hectare/ year)	**Total value** (per year)
Oceans	33,200,000,000	US$ 252	US$ 8,381,000,000,000
Coastal regions	3,102,000,000	US$ 4052	US$ 12,568,000,000,000
Forests	4,855,000,000	US$ 969	US$ 4,706,000,000,000
Grasslands	3,898,000,000	US$ 232	US$ 906,000,000,000
Wetlands	330,000,000	US$ 14,785	US$ 4,879,000,000,000
Lakes/rivers	200,000,000	US$ 8498	US$ 1,700,000,000,000
Cropland	1,400,000,000	US$ 92	US$ 128,000,000,000

The Costanza approach, and subsequent similar studies, brought to public attention the new discipline of environmental economics. Current environmental economics acknowledges that, as the old song goes, 'money makes the world go round'. But it points out that our current inability to sustainably manage biodiversity has to do with the workings of economic systems that do not account for the total value and usefulness of biodiversity.

As you can imagine, Costanza's 1997 article caused a storm after it was published, and continues to be debated today. There are problems with calculations, assumption, what you include, what you exclude. But the bottom line is that, warts and all, this was the first serious attempt to value the earth, and in a currency which those who are major users of biodiversity understand. However, as the authors freely acknowledge, all their calculations ignore the fact that many ecosystem services are 'literally irreplaceable'. And without such ecosystem services there would be no direct-use value.

how biosphere 1 works – as one

David Bowie asked the question, and H. G. Wells gave an imaginative answer to the same question: Is there life on Mars? 'Spirit' rover arrived on the red planet in January 2004 and is still operating as I write this. Despite 'Spirit' being the latest in a number of Martian landings, we still have no definitive evidence for extraterrestrial life. A group of scientists led by David McKay of NASA's Johnson Space Centre published an article in the magazine *Science* in 1996 in which they claimed to have found evidence for primitive bacterial life on Mars. This was in the form of microfossils and altered chemistry from a 1.9 kg meteorite, ALH84001, that landed in Antarctica 13,000 years ago. Experts in palaeontology were doubtful that the impressions on the rock were fossil bacteria, but there is no doubt that looking for chemical signs of life's activities is certainly one way to answer Bowie's question. Accounts published in early 2005 of methane production as a chemical sign of life (possibly bacteria) on the Martian surface have recently been superseded by a study which shows that methane probably just comes from Martian rocks. Still, life modifies non-living material and that is one way of gathering evidence for extraterrestrial life. It is also a way of studying the influence biodiversity has on the planet with which it is intimately associated, even if that planet is just our own one.

Compare the chemistry of the atmosphere of three planets that can be found relatively close to one another – Venus, our earth and Mars (Table 4). The atmosphere of Venus and Mars is nearly all carbon dioxide, with a little nitrogen but next to no oxygen or methane. If we were able to look at the earth without biodiversity we would find a similar story. But with biodiversity, most of the atmosphere is nitrogen, oxygen makes up about one-fifth of the atmosphere from plants and the like, and there is an accumulation of methane from bacterial activity. Carbon dioxide is now found in trace amounts. Even though it is produced when many life forms respire, it is also used up in the process of fixing the sun's energy by photosynthesis. In other words, the presence of life modifies the environment of our planet in a way that has not been found on any other planet.

Table 4 Comparing the gases in the atmosphere of Venus, the earth (with and without biodiversity) and Mars

Atmospheric gas	Venus	Earth (+biodiversity)	Earth (minus biodiversity)	Mars
Carbon dioxide	96.5%	0.03%	98%	95%
Nitrogen	3.5%	79%	1.9%	2.7%
Oxygen	trace amounts	21%	none	0.13%
Methane	none	trace amounts	none	none

The notion that there is an intimate, ongoing relationship between biodiversity and the physical make-up of our planet can be found in the writings of a Russian scientist from the late 1920s, Vladimir Vernadsky (1863–1945). He wrote that: 'Life appears as a great, permanent and continuous infringer on the chemical "dead-hardness" of our planet's surface ... it [life] is intimately related to the constitution of the earth's crust, forms part of its mechanism, and performs in this mechanism functions of paramount importance, without which it would not be able to exist.' Such a view, that to understand the workings of our planet you had to take account, and understand the role, of biodiversity, popped up again and again in scientific writings but never really caught on. The concept of the earth itself being, in some way, a superorganism, alive, only began to be taken seriously principally through the work of a British scientist.

James Lovelock was employed by NASA to design probes to detect life on Mars in the 1960s.[www#30] In his quest to detect life, he started first with the question, 'What is it that I'm trying to detect?' He settled on the fact that all living things take in energy, in the form of light, chemicals or food. They also form and get rid of waste products, and all of this they probably do using the planet's atmosphere as an exchange medium. If there were life on Mars this should, as we discussed above, be detectable as some sort of chemical signal. So Lovelock and another scientist, Dian Hitchcock, put together the sort of comparison of planetary atmospheres that we saw earlier. They concluded that while Mars was chemically 'dead', the earth was not – there were still major chemical reactions going on. Lovelock suggested that: 'The earth's atmosphere was an extraordinary and unstable mixture of gases, yet I knew that it was constant in composition over quite long periods of time. Could it be that life on earth not only made the atmosphere, but also regulated it – keeping it at a constant composition, and at a level favourable for organisms?' Lovelock outlined his idea to his friend William Golding, of *Lord of the Flies* fame, while the pair were out for a walk. Golding suggested the name Gaia for the idea, after the earth goddess of Greek mythology. So, in 1979, Lovelock published a book, *Gaia: A New Look at Life on Earth*, which spelt out the view that 'the physical and chemical condition of the surface of the earth, of the atmosphere, and of the oceans has been and is actively made fit and comfortable by the presence of life itself'.

While the idea that 'life' made the earth 'fit and comfortable' was an attractive one, exactly *how* life did this was still far from clear. An American microbiologist, Lynn Margulis, was investigating the processes by which living things (mainly microbes living in soils) added and removed gases to and from our atmosphere. Collaborating with Margulis, Lovelock put forward the idea that there were feedback loops which regulated aspects of our planet's environment. Can you remember science lessons in school where you may have tried to learn how a thermocouple worked? Well, Margulis and Lovelock proposed that the same principle was operating on a planetary scale.

Let's take one topical example. Carbon dioxide is the gas produced by many living things when they respire. But it is also produced in massive quantities beneath the planet's surface and released by such structures as volcanoes. Carbon dioxide is a greenhouse gas: that is, as its amount increases in the atmosphere, more heat is trapped and the earth warms up. Without some way of counteracting increases in carbon dioxide – without some feedback loop, if you like – our planet would continue to heat up until life was not possible. Certainly, living things can take up or give out carbon dioxide – e.g. plants can give it out and animals can take it in – but by and large these exchanges balance one another, so they are little help at mopping up additional carbon dioxide. Now, one of the ways in which carbon dioxide is removed from the atmosphere is through the weathering of rocks. Water, in the form of rain, and carbon dioxide from the atmosphere react with rocks to form chemicals known as carbonates (baking soda, sodium bicarbonate, is one; calcium carbonate, or chalk, is another). But vital to the regulation of carbon dioxide is the fact that this chemical process is greatly enhanced by the presence of bacteria in the soil. Importantly, the bacteria become more active as the temperature increases. So removal of carbon dioxide increases as the planet becomes hotter. The carbonates within which the carbon dioxide is locked are then washed into the seas and oceans where marine life forms use them to make shells. Most of these shells are tiny, belonging to microscopic plants that float in the oceans as plankton. When these shelled creatures die, the shells sink to the sea bottom and make up layers of what could one day become limestone. Some of the carbon dioxide trapped in these limestones will be released back into the atmosphere, as the rock may sink deep into the planet, where it melts, and the carbon dioxide is released through volcanoes, and so the cycle continues. Such carbon dioxide regulation is one of the many ecosystem processes or services we referred to earlier.

Criticism of Lovelock's ideas came from two main quarters. There were those who thought Lovelock was making a religious statement, claiming that the earth possessed 'a life force' actively regulating all of these services, while others, who defined life in terms of natural selection (i.e. the ability to evolve), could not see how superorganism earth could pass on its genes, so to speak. Lovelock's response to all of this was 'Daisy world'. He suggested that the earth may act as a superorganism and have its own 'physiology', its own ways of maintaining itself. Lovelock and a colleague, Andrew Watson, proposed the 'Daisy world model' as a way of trying to explain how the earth could do this. On an imaginary planet, which they call 'Daisy world', there are two different types, species, of daisy. Heat for Daisy world comes from a nearby sun, but the planet temperature is affected and ultimately controlled by these two species. There are black daisies. They increase in number, absorbing heat from the sun and warming the planet up (demonstrating why I should not have turned up for a job interview in Nevada in my black suit!). And there are white daisies. They do the opposite, increasing in number if the planet starts to get too hot. Between the changing numbers of the two species the planet's temperature remains roughly stable.

how 'bits' of biosphere 1 work

Gaia has been a useful hypothesis for trying to work out how our planet and its inhabitants (biosphere 1) work and interact. Noticing and predicting patterns is one thing, but actually carrying out experiments is another. And yet over the past fifteen years there has been increased interest in investigating experimentally in what way(s) biodiversity determines how ecosystems work. In particular, does manipulating the numbers and/or identities of species in a particular ecosystem affect the sorts of thing that ecosystem does, contributing to the overall function of the biosphere? Such questions are interesting in their own right. How many species can we afford to lose before ecosystem function fails? Are all species equally necessary for ecosystem function and, if not, which ones are? Of course much of what is driving non-scientific interests are the same sorts of questions but with a different spin. How many, and which, species do we need to protect and conserve to keep our life-support systems functioning?

But first things first. Why should we think that there will be a relationship between what biodiversity is and what it does? We already know that some species have a much greater influence on what ecosystems do than others. And the more species you have in your ecosystem, the more likelihood that you will have a greater number of 'influential' species. And even if species were equally 'influential', they may be influential in different ways and so complement, or even positively affect, each other. So there is some reason to suspect that there is a relationship between the biodiversity of an ecosystem and how that system works. That being the case, what sort of relationship could we expect?

A number of suggestions have been made as how best to describe the relationship linking the number of species in an ecosystem to the way that ecosystem works (Figure 12). It may be that how good the ecosystem is at doing what it does just gets better as you add more species; so an increase in species richness means increased ecosystem function. This is the diversity-stability hypothesis. The second suggestion is that a loss of species will not affect how the ecosystem works, at least down to a critical level. Those remaining can make up for the losses, although there is a minimum number of species needed to make the whole thing work. This is referred to as the redundancy hypothesis. The third suggestion is that losing a few species may seem to have little effect, but the more you lose the greater the detrimental effect on ecosystem function. This is termed the rivet-popping hypothesis, with the analogy being that species in an ecosystem are like rivets holding an aeroplane together. The fourth suggestion is that the world is a complicated place, species have numerous complicated roles

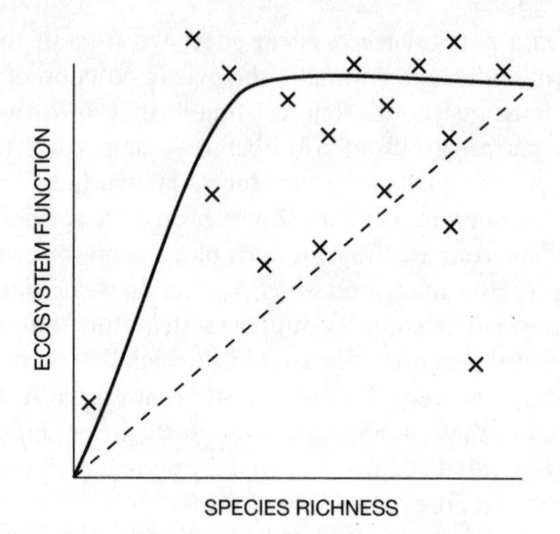

Figure 12 *Graphs of possible relationships between biodiversity and ecosystem function (solid line = redundancy hypothesis; broken line = diversity-stability hypothesis or, if instead of a straight line we make it stepped, the rivet hypothesis; crosses = idiosyncratic hypothesis)*

and interactions, so that while a change in biodiversity will result in changes in function, you don't have a snowball's chance in hell of putting forward one overall theory that will explain everything. Here we have the aptly named idiosyncratic hypothesis. That's not to say the relationship is necessarily unpredictable. For instance, it may be that the removal of one key species will have a different effect from removing other species in the ecosystem. A species with such a pivotal role in how an ecosystem is put together is called a keystone species.

And finally, despite what I said earlier about having good reason to expect a relationship, there is still the fifth possibility that there is just no relationship.

Where have we got to in testing these ideas, these hypotheses? Much of the experimental work has involved looking at plant (often grass) species richness and relating that to productivity. Dave Tilman and his team, for example, took a large number of fields where they manipulated the numbers of grass species present. BIODEPTH was a European project, with eight very different sites, that set out either to confirm or to refute the existence of a relationship between species richness and productivity in grasslands. In each case, and with all the studies that have been carried out to date, the results have been complicated. Interpreting the results has been made difficult by the way in which some of these very complicated experiments have been designed and there is considerable debate in the scientific literature as to what can be deduced from them.

One of my graduate students, Sally Marsh, and I carried out an experiment to investigate whether the number of different species of sandhopper present ('biodiversity') was related to decomposition of their food ('ecosystem function'), which is large bits of seaweed cast up at the high-tide mark. The greater the number of species present, the greater was the decomposition, even though the number of individuals used and their body masses were standardized. These were three species which, on the surface, all look as if they do the same thing in the ecosystem – break down seaweed. And yet loss of one or two species affects the way the ecosystem works. Here we have a simple, clear-cut example of how even the diversity of what appear to be very similar species makes a huge difference to ecosystem function.

Many think the period of investigating biodiversity and ecosystem function in the way described above is coming to an end. In some ways our understanding has advanced. It does seem, overall, that if you increase the total number of species present, you increase the number of redundant species. But this is far from the whole story. It is complicated by the fact that changing the conditions within an ecosystem may change the identity of the 'influential' species present. So it is not just how many species make up an ecosystem that is the issue, but instead making sure that the ecosystem can draw on a pool of species to act as a sort of buffer to change. You will have gathered by now that we are still a long way from identifying some grand unifying theory. Taking a broad brush approach, all of the experiments carried out to date have only confirmed what biologists have known for years and rural communities for centuries – greater biodiversity is better for the functioning and stability of ecosystems.

Another complementary approach to understanding how ecosystems work is to attempt to build your own. This is what we will explore next.

build your own biosphere – not so silent running

the home marine aquarium

A couple of years ago I had a few hours to kill in Charlotte Airport in the US. In one of the airport shops I found a small hand-blown glass ball for sale. Completely sealed off from the outside world it contained a couple of tiny red shrimps, a beautiful piece of gorgonian coral and some microbes, all living in seawater. This was retailing as an executive toy, the Ecosphere. All it needed from outside was a certain amount of sunlight. A spin-off from NASA research into sustainable ecosystems in outer space, this was a self-sustaining system that would last for about three years. The associated blurb claimed that some could last as long as ten. A world in the palm of your hand.

One of the first serious attempts to construct a self-supporting ecosystem was born from the idea of having a low-maintenance aquarium in English mid-nineteenth-century middle-class homes. A certain gentleman, Mr Robert Warrington, presented some early experiments to the British Chemical Society in 1850. His scientific paper was entitled 'On the adjustments of the relations between animal and vegetable Kingdoms, by which the vital functions of both are permanently maintained'. He told the assembled gathering that he placed two goldfish in twenty gallons of spring water which half-filled a glass bowl. Some sand, mud, pebbles and bits of limestone were carefully added to carpet the bottom of the bowl. The mud was used to hold the roots of a small water plant called *Vallisneria.* Warrington then left the aquarium undisturbed until a green scum started to coat the walls of the bowl, obscuring his view of the goldfish, and the goldfishes' view of him. The water began to cloud over and look distinctly unhealthy. The addition of some snails, which fed on both the green scum and some of the decaying vegetable matter, resulted in the water clearing in a very short space of time. Minor pruning by Warrington kept the plants happy and healthy, the snails flourished on shed leaves and the fish grew fat and healthy on the snails' eggs for many, many months.

A mere three years later he had published a number of essays on how to set up a self-sustaining marine aquarium. This coincided with the beginning of a Victorian love affair with the seaside. Numerous natural historians, including Phillip Henry Gosse, of *Father and Son* fame, got involved in producing what were in effect self-sustaining ecosystems. Gosse was drawn into the construction of London's first large-scale public aquarium, in the Gardens of the Zoological Society. Through books such as his *Aquarium* and *Devonshire Rambles* he sparked off a craze in keeping long-term marine aquaria in private houses, often many miles from the sea. This was the first commercial attempt to sell the idea of self-sustaining ecosystems (or closed ecological systems, as they have come to be known).

Using closed ecological systems as a research tool has largely been overlooked since those days of the first marine aquaria. The beginning of serious research can be traced to the early 1960s and has evolved from the use of small glass bottles right through to the Biosphere 2 project and the NASA-funded Closed Ecological Life Support System (CELSS) project.

mysteries and hazards

Silent Running (1972) was a cult science-fiction movie which had Freeman Lowell, a nature-loving astronaut, aboard the *Valley Forge*, a gigantic spaceship in a small fleet that carried the last surviving forests of an overpopulated and environmentally degraded earth. The film ended with three little pre-R2D2 robot characters tending, watering and generally looking after the forests as they floated endlessly through space. Here was a self-sustaining ecosystem, a little biosphere that would have been capable of sustaining Lowell and the other three crew, had Lowell not 'disposed' of his companions early on in the film when they wanted to dump the forests and return to earth. Keeping a bunch of marine animals and plants healthy for a couple of months or even years is one thing, but building an ecosystem that can support human life is another altogether. How close are we to science fiction becoming science fact?

Biosphere 2 (remember the earth is biosphere 1) was an airtight, 204,000 cubic metre facility.[www#31] Seven years in construction, it was completed in 1991 at an estimated cost of US$200 million. Lots of small bits of large natural systems – coral reefs and grasslands, for example – were crammed together in what looked, from a distance, like a huge greenhouse. The 'experiment' consisted of enclosing eight people inside and seeing if the human-made biological system could supply their food, water and air needs for two years (1991–93). It is difficult to know how many of the 'challenges' to the lives of the biospherians, as they were called, could have been predicted. The microbes in the soil that was used to grow food consumed much more oxygen and produced much more carbon dioxide than was expected. A good proportion of the carbon dioxide reacted chemically with calcium-containing concrete used in the construction of the biosphere. The chemical result of this reaction was chalk. The atmospheric result was that a lot of the oxygen contained in carbon dioxide that was needed in order to be transformed back into breathable oxygen by the plants was, in fact, locked tight in the structure of the building. So within a year and a half of the 'lock-in', breathable oxygen had fallen to two-thirds of its normal level (equivalent to living at the top of a very high mountain) and carbon dioxide was high and all over the place. Just in case you're holding your breath by this time, don't worry – before the two years were up, the outside 'back-up' team added oxygen to the biosphere atmosphere from outside. But even this did not mean that the biospherians could breathe easily, because the plants introduced to recycle 'waste' carbon dioxide had their own surprises. Some of the vines turned out to be very competitive. Vigorous weed management failed to stop species like 'Morning glory' threatening huge areas laid aside for food production. Bits, often big bits, fell off big trees, nineteen of the twenty-five (non-human) vertebrates went extinct. So too did many of the insects and all the pollinators. Plants that rely on pollinators for reproduction were technically extinct – they just hadn't caught on to the fact. Wonderfully named crazy ants did really well, as did a few cockroaches. Aquatic environments became polluted because of nutrient overload – just like biosphere 1.

The bottom line was that the eight people within Biosphere 2 were quite literally involved in a life-or-death struggle every day – a struggle to maintain services from nature that we enjoy daily, take for granted, and to a large extent get for free. Despite almost unlimited back-up from the outside, it was impossible to maintain a closed system that supplied the material needs of just eight individuals for two years. In the words of Joel Cohen and Dave Tilman, assessing the venture (*Science*, 274, 1150–1),

> At present there is no demonstrated alternative to maintaining the viability of earth. No one yet knows how to engineer systems that provide humans with the life-supporting services that natural ecosystems produce for free ... Despite its mysteries and hazards, planet earth remains the only known home that can sustain life.

valuable for what, and to whom?

The whole notion of non-use value was first put forward in 1967 by John V. Krutilla of Resources for the Future, a nonpartisan think tank based in Washington DC. Although differing according to what book you read, there are essentially three non-use values: option, bequest and existence (Figure 11).

keeping options open

Option (or potential) value has been described as the value of something that has not yet been recognized or, as Joni Mitchell put it more poetically if negatively: 'You don't know what you've got 'til it's gone/You pave paradise and put up a parking lot.' Once biodiversity is lost, the option of benefiting from it, even if you don't currently know what these benefits might be, is gone for ever. In moneyspeak, option value is like an insurance premium that you are willing to pay to retain the option of future use on a good or service whose future you're unsure of. One of the best illustrations of option value is the International Rice Research Institute's seed bank collection, set up in 1961.

I've already mentioned rice as one of the most important food plants on the planet. It provides food for more than half the world's population. In the 1970s rice crops were infected by the grassy stunt virus, transmitted by the brown plant hopper which infested huge rice-growing areas in Asia. The high-yield varieties of rice in use had no resistance to the virus. The International Rice Research Institute maintained a seed bank representing many thousands of rice varieties, including recently developed hybrids and strains, as well as wild varieties that were no longer used in cultivation. They tested in excess of 6000 different rices for resistance to the virus. Only one, a strain no longer in use, was resistant. In 1976 this redundant strain was crossbred with others and a new strain created, known as IR36. IR36 is both a high-yield rice and is resistant to the virus.

The message from this story is that without access to a rice strain thought to be redundant, the economies and lives of many, many people would have been quite different, and possibly negatively so. In this case, and in many hundreds of others, genetic information was viewed as a source of novelty, with extinction literally being an irrevocable loss of that information.

bequesting and bequeathing

Scottish philosopher/economist John Locke was one of the first to say that 'each generation should bequeath enough and as good for others to future generations as justice demands it'. This is termed bequest value. A commonly given example is that of preserving a national park. Even though not all of the present generation will ever have the intention of visiting the park let alone use and enjoy it, its preservation will benefit future generations. But bequest value is much more than just leaving something behind for future generations. It encompasses the notion of compensating future generations for the loss of biodiversity value – be it money, production or ecosystem services – we are responsible for right now, recompensing our children and their children for our use of what should have been their biodiversity. For example, economists sometimes use surveys to estimate the monetary value of bequest value by asking questions such as how much money would people be willing to pay to save, say, the tiger or the panda, or a piece of land for future generations.

just being there

There is one more set of non-use values that we need to consider: existence and intrinsic. These are notoriously difficult to disentangle and describe. Existence value could be defined as the value we, as people, place on just knowing that a species, a habitat, or an ecosystem exists, even if we will never see it or use it. To convert this intangible value into hard cash we could go back to our economists with their survey and ask people how much money they would be willing to pay just for the knowledge that the tiger, the panda or the piece of wilderness exists (existence value). This is one, but only one, way of doing it. Part of the problem with the approach, whether we are talking about bequest or existence value, is that such questions are always hypothetical. (If you had to, how would you murder your wife? Well, I would ...)

For the remainder of this chapter we will trace the history of some of the ideas, philosophical and religious, that have touched on intrinsic and existence value and see how those ideas have shaped our present-day 'world views' of biodiversity.

full-on philosophers and laid-back religion?

value bestowed, not intrinsic

Throughout our history, biodiversity, and nature generally, has furnished us with the raw materials and the inspiration for our existence and our lives. It should not be surprising, therefore, that much of the value of biodiversity has been tied up with our own existence, well-being, cultural activities and spiritual lives. It was perhaps summed up most succinctly by the Scottish philosopher John Locke, who claimed that everything in nature was waste until we transformed it into things of value. There is not necessarily any intrinsic value – it is only valuable to the extent that it satisfies our physical, mental and spiritual needs. Arguably, this has been the most influential idea and assumption, dominating Western thought from classical times. The Greek tragedian Sophocles (496–406 BC), the polymath Aristotle (384–322 BC) and Roman stoic Epictetus (AD 55–135) all propounded ideas on the excellence of human nature and the precedence of humans over all other species. Sophocles even went as far to say that humankind 'is master of ageless earth, to his own will bending ... He is Lord of living things.' This belief in human superiority and domination of nature was a common theme in medieval and Renaissance thought, despite a few dissenters. The Church taught that God's commands in the book of Genesis, to dominate and subdue, meant that nature was there for us to do as we will. We were its masters. This is clearly seen in Thomas Aquinas's *Summa Contra Gentiles*, where he says that: 'Hereby is refuted the error of those who say it is sinful for a man to kill dumb animals: for by divine providence they are intended for man's use in the natural order. Hence it is no wrong for man to make use of them, either by killing or in any other way whatever.'

Human superiority and domination were no less prevalent in secular thinking, and to some extent made easier by the views emerging from Renaissance scientists and thinkers, like Descartes and Galileo, of nature as a machine. Conquest and subjugation of nature in different forms continued to dominate thoughts on our relationship with biodiversity right up to the present day, even if ideas such as Darwin's theory of natural selection, and evolution generally, have done much to show our own 'less exalted' place in the natural order. Humans had become more thoroughly part of biodiversity, but this did not stop them lording over it. Some philosophers, like the German Immanuel Kant (1724–1804), in some ways embodied the prevalent view that nature had no purpose, so he suggested that we act accordingly. In the twentieth century, philosopher Jean-Paul Sartre (1905–80), biologist Jacques Monod (1910–76) and philosopher Bertrand Russell (1872–1970) all held the view that nature has no value except that which we project on to it. Perhaps the most extreme view was that of Friedrich Nietzsche (1844–1900), who tried to destroy the ideas of nature as a machine, as a living being, as having any purpose or any beauty and harmony: 'The total nature of the world is ... to all eternity chaos. The living being is only a species of the dead, and a very rare species.' Much of human history has seen the purpose of economic growth and technological advancement as maintaining and enhancing our material well-being. Thus, the whole idea of conservation in this world view is one of efficient management of limited resources, converting wild and hostile nature into a more benign, more 'comfortable' environment.

intrinsic value

Petrarch (1304–74) believed that nature was a sign of God's providence, but he also thought that it existed for its own sake. In his book *The Ascent of Mont Ventoux*, he writes: 'Today I ascended the highest mountain of this region ... I admired every detail.' German explorer Alexander von Humboldt (1769–1859) declared his excitement and aesthetic delight when encountering the natural world. Aldo Leopold, an American naturalist and the 'father' of the study of wildlife ecology, was one of the most famous supporters of the idea that wildlife and wildlands were valuable in and of themselves. He wrote:

> The last word in ignorance is the man who says of an animal or plant: 'What good is it?' If the land mechanism as a whole is good, then every part is good, whether we understand it or not. If the biota, in the course of aeons, has built something we like but do not understand, then who but a fool would discard seemingly useless parts? To keep every cog and wheel is the first precaution of intelligent tinkering.

Those who would question the existence of any notion of intrinsic value would point out that Petrarch and all the others who say they enjoy nature for its own sake still have the major emphasis on their enjoyment and not the intrinsic value. Despite this type of objection, we still have reference to the intrinsic worth of biodiversity contained in many of the treaties and conventions drawn up over the past decades, using such value as one of the bases for conservation. Deep ecology is a world view that has emerged over the past few years, which has at its centre the recognition of nature's right to exist quite apart from any benefit we may derive from it. So, despite the fact that it is difficult to demonstrate, the view that biodiversity has intrinsic value has been relatively widespread over the last hundred years or so. However, even those who hold that biodiversity and nature have their own right to exist, and should be conserved for their own sake, acknowledge that our own future and that of biodiversity are inextricably intertwined.

valued as an object of worship or through kinship

The worship of nature, which includes biodiversity, is referred to as pantheism. Pantheism is a persistent and all-pervasive feature of human history and belief. In the oldest Hindu scriptures, the Vedas, which are three-and-a-half thousand years old, nature is worshipped and revered. The influential Portuguese-Jewish philosopher Spinoza (1632–77) believed in the divinization of nature. Today, such divinization is found in many forms, from old-established religions, such as Paganism, through to 'new age' splinter groups. There is a version of the Gaia hypothesis where Gaia as a biological theory imperceptibly metamorphoses into the earth mother of mythology. There are even 'Christian' forms, such as the Process Theology of A. N. Whitehead (1861–1947), where God is evolving as part of his creation

Perhaps more common than the divinization of nature is the world view that humans are very much an 'equal' part of biodiversity. This is undeniably true in a biological sense – all life is related – but the idea that animals in particular are our kin takes this a stage further, claiming an even greater unity. Taoism, as embodied in the writings of the mystic Lao-Tze (sixth century BC), emphasized the essential unity of humanity and nature. The same sort of belief was held by many indigenous peoples such as native Americans. The Lakota nation believed that 'all things were kindred and brought together by the great mystery'. Schopenhauer (1788–1860) was one of the first modern philosophers to attack the notion that humankind is better than nature. The Scottish-American naturalist John Muir (1838–1914) believed in a harmony between humankind and nature, spiritual as well as physical. This he shared with other American writers and thinkers of the nineteenth century, such Ralph Waldo Emerson and Henry Thoreau. Muir loved wilderness and was a huge influence in promoting the idea of forestry conservation and setting up national parks. In many ways he is seen as the first person to develop a modern environmental ethic. In *The Wild Parks and Forest Reservations of the West* he wrote: 'Thousands of tired, nerve-shaken, over-civilized people are beginning to find out that going to the mountains is going home; that wilderness is a necessity and that mountain parks and reservations are useful not only as fountains of timber and irrigating rivers but as fountains of life.' The modern version of this unity or kinship idea is what is known as the Biophilia hypothesis.

Harvard biologist Edward O. Wilson is regarded by many as the father of modern biodiversity and has perhaps done more than anyone else to bring our biodiversity crisis to popular attention. He published a slightly different type of book in 1984, entitled *Biophilia: The human bond with other species.* Here he put forward his belief that we human beings, as part of the natural world for many hundreds of thousands of years, have a natural, innate, regard and need for living things. This he called Biophilia, which he defined as 'the connections that human beings subconsciously seek with the rest of life'. Other advocates of this emotional or spiritual dimension to our relationship with biodiversity as evidence for Biophilia, point, for example, to studies that have shown that patients recovered quicker if they were exposed to greenery, even images of greenery, compared to recovery in an artificial environment. Some have even linked Biophilia with the Gaia hypothesis (humans are one element of Gaia) and, not infrequently, such discussions border on the divination of nature that we started with.

a creator gives biodiversity value

While it is true that for much of our history there has been a religious as well as a secular emphasis on the superiority of humankind and the value of nature solely in terms of satisfying human needs and desires, the reasons for this are not always as straightforward as is claimed by many modern commentators. Much of the blame for the overexploitation of biodiversity, and nature in general, is laid at the door of religious belief. In particular, those faiths that seem to draw a firm distinction between humans and the rest of bio-diversity – Judaism, Christianity and Islam – come in for criticism as being biodiversity-unfriendly, while belief systems like Hinduism, Buddhism and Taoism are held to be more biodiversity-friendly. However, there are two points that should be made. First, whether a country or a culture is characterized by a biodiversity-friendly or allegedly biodiversity-unfriendly belief system is not clearly reflected in the ways that these countries or cultures, in the past or today, treat biodiversity. The widely held view that peoples whose religions require them to live in close contact with nature and to respect it have not been responsible for major historical extinctions has very little factual basis. The truth is that most human beings, irrespective of colour or creed, throughout their history have treated biodiversity – on different scales, admittedly – rather badly whether they were aware of it or not.

Second, just because Judaism, Christianity and Islam hold that there is a qualitative difference between humans and the rest of biodiversity, it does not necessarily follow that they should treat biodiversity badly. That may happen, but not as a result of the beliefs of those faiths. In fact, the very opposite should be true. The whole tone of the creation account in the book of Genesis (and many other sections of what Christians refer to as the Old Testament) is one of stewardship of the earth and its resources. Man (or woman) is not the measure of all biodiversity. While humans are considered different (and special) from biodiversity as a result of individually bearing 'the image of God', these writings also make it clear that we were created from the same stuff (dust, earth) and eat and reproduce as the animals do. This is what led Blaise Pascal to emphasize the humble place of humankind at a time (the Renaissance) when both religious and secular thought seemed bent upon the 'divinization of man'. Not only are humans *not* the master of biodiversity in these sacred writings, but in fact they are portrayed as being charged with looking after the planet on behalf of someone else – and that someone else is the creator of the planet and all that is on it. We are depicted as stewards – not owners – and God is the creator, sustainer, the owner of biodiversity. In the management process humankind were granted leave to use biodiversity for their own subsistence, but this is a far cry from Aquinas's 'do whatever you want with biodiversity'. The English translations of the book of Genesis which talk about 'having dominion over' and 'subduing' the earth convey a very different sense from the stewardship theme that runs through the scriptures of these three world faiths. In the Jewish scriptures, the Christian Old Testament, there are laws about the treatment of land and biodiversity. In one of them, soldiers were forbidden to cut down fruit trees even if their lives were in danger. Another theme running through the beliefs of all three faiths is the view that biodiversity and the natural world glorifies and reveals God. Although this notion has a long religious history, it was actually brought to bear on scientific study, as its motive and impetus, from the Renaissance up until Darwin's time. Both John Ray (1627–1705) and William Paley (1743–1805) both thought nature worthy of study and attention because it revealed God. The argument from (or to) design is still hotly debated to this day.

One of the most influential Christian theologians of the twentieth century, Karl Barth (1886–1968), believed that we had a moral responsibility for nature. Barth was very wary of any notion of 'reverence for life' but believed that,

> the world of animals and plants form the indispensable living background to the living-space divinely allotted to man and placed under his control. As they live, so can he. He is not set up

> as lord over the earth, but as lord on the earth which is already furnished with these creatures. Animals and plants do not belong to him; they and the whole earth can belong only to God. But he takes precedence of them.

This is quite a different emphasis from what seems to be the case in right-wing American Christianity at the moment, and it also stands in sharp contrast to the practice of other Christian groups throughout the world. There are, however, indications that this is changing. Christians who are also eminent biologists, such as Professors Sam Berry and Sir Ghillean Prance, have been very proactive in recent years in raising the awareness of the biodiversity crisis among mainline Christian groups (e.g. by the formation of 'An evangelical declaration on the care of creation').

to conclude

What is very clear is that biodiversity has value, and value over and above its direct-use values. This value may be difficult to categorize and pin down, but few would deny the concept of 'worth'. The prevalent view is that, from either a religious or a secular point of view, with different motivations, we have a moral responsibility to protect the life forms with which we share this planet. With this in mind, we move on now to look at the steps we, as a global community, have taken to look after this biodiversity that, in theory at least, we value so much.

chapter seven

leaving the world as you would like to find it

'It is time you lavished on your rose which makes your rose so important.' 'It is time I lavished on my rose ...' said the little prince. So as to be sure to remember. 'Men have forgotten this basic truth' said the fox. 'But you must not forget it. For what you have tamed, you become responsible forever. You are responsible for your rose ...'

'I am responsible for my rose ...' the little prince repeated, so as to be sure to remember.

The Little Prince, Antoine de Saint Exupéry

saving private land

I open the latest letter from the National Trust, of which we have family membership. It's now official: Wembury Point, an ex-Ministry of Defence site immediately to the west of Wembury Bay, has been saved from property developers. It is difficult to believe that this specially designated Site of Special Scientific Interest (SSSI), recognized nationally as an area of outstanding beauty and a hotspot of biodiversity, was under threat in the first place. But it took major intervention by the heritage watch-dog English Nature to make sure that the Ministry of Defence offered the site to a conservation organization before placing it on the open market.

It's a conservation success story and one, for me, very close to home. And there are hundreds, if not thousands, of such stories worldwide. But it would be folly to believe that Wembury Bay and its surroundings are now safe. What happens in Plymouth, in Devon, in the seas surrounding the bay, in this part of Europe, globally, all affect efforts to conserve the biodiversity of Wembury Bay. Without action on an international and global scale, our very best efforts to protect the biodiversity of Wembury could turn out to be as significant as rearranging deckchairs on the *Titanic*.

The subject of biodiversity was born out of the need to gather information for conservation purposes. So the conservation of biodiversity is a fitting way to end this book. We will attempt to trace the main global, international and national conservation issues of the past and assess their state now, in the present. As already alluded to, we could pack pages and pages with stories of local biodiversity conservation success stories (or probably an equal number of 'failures'), but given the global nature of biodiversity, this global level is what we will deal with here. That is not to say that the local stuff is not important. Without it, perhaps, we would have no biodiversity to protect, internationally or globally. 'Either it is all important or none of it is.'

antecedents

Although the origins of the modern conservation movement can be traced back to the middle of the nineteenth century, it was only after the Second World War in the twentieth century that we begin to pick up global concerns about global wildlife. In 1972 the United Nations convened the first global conference ever to address environmental issues. That meeting stimulated the formation of environmental ministries and civil organizations promoting environmental issues throughout the world. The United Nations Environment Programme (UNEP for short) was established.[www#32] There was the explicit acknowledgment that the forces driving the destruction of the natural world lay outside the remit of biology, and so biology alone would not be enough to achieve some degree of protection. Integrating science into the bigger picture that included politics, economics, morality, social responsibility and so on was seen as key to conservation at every level.

The abbreviated form of biological diversity, biodiversity, was coined by Walter Rosen, the co-director of a conference in 1986 entitled 'The National Forum for Biodiversity'. The book that resulted from this conference was *BioDiversity* (1988), edited by E. O. Wilson.[www#33] To a large extent this book sparked off both the concept and the use of the word 'biodiversity' in a way that continues even today. Wilson's beautifully written, more populist book *The Diversity of Life* (1992) did much to bring biodiversity into popular thought and consciousness. Pressure from UNEP and a large number of environmental groups, together with a growing awareness of biodiversity and its loss in the mind of the public, resulted in the first global attempt to conserve biodiversity in Rio de Janeiro at the beginning of the 1990s.

oh rio

The Convention on Biological Diversity (CBD) was a landmark document in our attempts to look after biodiversity.[www#1] Signed by more than 150 nations on 5 June 1992, the treaty arose from the United Nations Conference on Environment and Development, held in Rio de Janeiro, Brazil. It came into force about eighteen months later. On the plus side, and not to be underestimated, this was the first attempt at a truly global treaty, and emphasized more than ever before that the conservation of biodiversity was a common concern, no matter who you were or where you lived. Genetic diversity was included explicitly for the first time as a conservation priority. On the downside, many countries were slow to ratify the document and even now a good number of the key players have yet to come to the party.

Despite its complicated legal jargon and multiple caveats, the ideas that are core to the treaty are relatively easy to grasp. The objectives of the CBD are: (1) the conservation of biological diversity; (2) sustainable use of its components; and (3) equitable sharing of the benefits arising from the utilization of genetic resources. The bottom line is that conservation, sustainable use and sharing of benefits must take place if only because otherwise we imperil our own existence. There are forty-two articles in the CBD. They range from carefully worded aims, objectives and definitions, through to policies that need to be implemented and a spelling out of what the signatories have actually signed up to. Overall, it's a lot more than setting up a few nature reserves and bringing a couple of cute furry or feathered creatures back from the edge of extinction. In what follows we will go through the main points of the CBD and also, to some extent, use this as a framework for introducing basic conservation concepts such as *in situ* and *ex situ* conservation.

large brush strokes

The first article gives the broad-brush picture of what the whole convention is about, conserving biodiversity, using its constituent parts in a sustainable way, and doing so in the context of 'fair and equitable sharing'. There's biology here, yes. But set firmly in a social, political and global context. All very laudable. But how is this to be achieved? Conserving biodiversity is not going to happen purely because a hundred or so countries sign up to a treaty that says that biodiversity is a good thing. As we have seen, there is little encouragement from our track record on how we have looked after biodiversity to date. As the British psychologist Havelock Ellis said: 'The sun, the moon and the stars would have disappeared long ago had they happened to be within the reach of predatory human hands.' So signature nations of the CBD were obliged to put into place strategies, tangible ways, by which biodiversity can be conserved and used in a sustainable fashion (Article 6).

louder than words

Article 6 is arguably the most far-reaching and significant article. For it to be effective, sustainability must touch on and include almost every area of a nation's activities. Let us be clear about this. Even if only Article 6 on sustainability was taken just semi-seriously it would fundamentally change the way nations are governed. One of the more obvious things that would need to change are perverse subsidies – financial support given to processes that in the long run have an adverse effect on both biodiversity and the economy. Support for fossil fuels, which increase pollution, negatively affect human and animal health and contribute to global warming, is a good example of a perverse subsidy. Some economists have estimated that sometimes the amount of money financing perverse subsidies often exceeds the marketable value of the goods being generated.

Article 6 asks that strategies and plans be drawn up that will affect biodiversity conservation on the ground. A tangible way we can see how this works is the article's effect on the management of a national park just north-east of Wembury Bay, Dartmoor National Park. As a result of the Rio summit in 1992 all the signatory nations had to put national action plans into place, documents that identified at the national level biodiversity priorities. The UK Biodiversity Action Plan[www#34] was launched towards the end of 1994, followed shortly afterwards, in 1995, by the setting up of the UK Biodiversity Steering Group. There followed the production of a series of habitat and action plans for different parts of the country. *Action for Biodiversity in the South West*, published at the beginning of 1997, started the planning process for a number of counties, including Devon. About a year later, *The Nature of Devon: A Biodiversity Action Plan* took the planning at local level a stage further, while English Nature and Dartmoor National Park Authority produced *The Nature of Dartmoor: A Biodiversity Profile.* By early 2001 *The Dartmoor Biodiversity Action Plan* was produced and by September of that year the Dartmoor Biodiversity Project was launched. Similar implementations of the CBD are going on all over the world, admittedly with different degrees of success and/or haste, but still they are happening. Depressingly, many such undertakings are often mere aspirations. Regularly there appears to be a failure to recognize the fundamental nature of exactly what needs to be done.

Putting biodiversity action plans and so on in place, exactly as we've seen above, is all fine and good, but how do we know that we are even close to achieving what we set out to do? The CBD 'recommends' a carefully thought-out programme of monitoring, although it does not say how this is to be done (Article 7). In effect, this is an information-gathering exercise, an audit of biodiversity and of the mechanisms put in place to conserve it and use it sustainably. We've already seen how difficult it is to get an idea of just the number of species on earth, never mind anything more complicated. This degree of realism is shared by the CBD, which suggests that such information gathering should be targeted on components of biodiversity believed to be key for conservation and sustainable use, and threats to those components.

arks in parks

One of the longer and tortuously complicated articles is Article 8 on *in situ* conservation – conserving things 'where they are'. This obliges signatories to set aside a number of protected areas within their national boundaries (sections a and b), to protect the land that borders protected areas (sections c, d and e), to attempt to restore or aid the recovery of degraded habitats (section f) and to combat the risks associated with alien species (sections g and h). I'll say more about the establishment of protected areas later (and one that didn't make it), but in the meantime we should note, in passing, that difficulties in providing financial and other support for *in situ* conservation, particularly to developing countries, is a recurrent theme of conservation. Sections i and j urge signatories to look for ways of minimizing conflicts between conservation and present use, whereas the final sections (k, l and m) highlight the legal and financial obligations of such conservation, recognizing that poorer countries will need help.

out of place, but alive

As well as conserving things 'where they are' (*in situ*), allowance is also made in the CBD for conserving species outside their natural habitats (*ex situ* conservation) (Article 9). While *ex situ* is always seen as second best to *in situ*, even second best is sometimes as good as it gets. Currently there are about 1500 botanic gardens and between 800 and 10,000 zoological parks (depending on how you define them) worldwide. They could be the key to *ex situ* conservation in terms of breeding, holding and reintroducing endangered species. There are also more specialized biodiversity 'holding facilities', such as sperm banks, tissue cultures and seed banks (such as the International Rice Research Institute's seed bank that we met in the previous chapter). The case for zoos as effective places for the conservation of biodiversity is still hotly debated. It has been pointed out that of the million or so individual vertebrates held in zoos very few of them have any great conservation status. In the case of mammals, it's been estimated at about one in ten. However, supporters of zoos point out that there are more than 300 endangered species in captivity, and zoos have successfully preserved species such as the Bison and Prejwaski's horse from total extinction. Certainly the earliest *ex situ* success story must be Père David's deer, which went extinct in the wild in China 3000 years before the present, but survived in an area of parkland. *Ex situ* is often seen as complementary to *in situ* conservation as it is sometimes possible to take species from captivity and reintroduce them back into the wild.

buzzword for the twenty-first century

Alongside biodiversity 'sustainability' must rank as one of the key buzzwords of this century. The sustainable use of the components of biodiversity is what Article 10 is all about. Given that, as we've seen throughout this book, most uses of biodiversity have not been (and are not) sustainable and that currently human exploitation is estimated to be 20% greater than the earth's productivity, this notion has an air of urgency about it.

Sustainable development is all about reconciling economic development with environmental protection. It's meeting the needs of the present, with an eye on protecting our natural heritage for our children and their children. For example, in the second half of the twentieth century cultivation of high-yield crops, with the intensive use of fertilizer and pesticides, resulted in considerable increases in productivity. More crops for each buck you spent. But putting values to many of the hidden costs showed that this was not such a good deal as we thought we were getting. The environmental and biodiversity costs were substantial. What was needed was a compromise between what was truly economically viable, what was technically possible and what was biologically acceptable. Today, taking sustainability issues more into account, cultivation systems are becoming more diversified, as is what farmers do and how they do it.

For our activities to be truly sustainable there has to be a major cultural shift. And this requires the support of local peoples, wherever they are. It is not just a case of going back to some 'golden age' when we allegedly 'lived in harmony' (sustainably) with nature. As already seen in previous chapters, primitive humankind had a negative impact on biodiversity, and even where there were examples of 'living in harmony' they only worked because of low human density and the absence of commercial exploitation. There is no golden age to go back to, and even if there was, we couldn't go back now anyway.

Article 11 is all about incentive measures. Signatories should adopt economically and socially sound measures that act as incentives for conservation/sustainable use. As biodiversity loss is driven by economics, we should harness the same forces to protect biodiversity. However, in reality very often the converse is true, as the abundance of perverse subsidies show. The remaining articles of the convention deal largely with procedural matters.

responses to rio

One of the potentially major outcomes of Rio was Agenda 21. Basically, in 1992 the international community signed up for what was billed as an unprecedented global plan of action for sustainable development. This meant both improving people's lives and conserving natural resources against the backdrop of huge population growth and the accompanying increased demands for economic security, energy, food, health care, shelter, sanitation and water. To this end, the UN Commission for Sustainable Development was set up in 1992.[www#35] Its job was to monitor and provide feedback on how Agenda 21 was being implemented on the ground at local, regional and international levels.

ten success stories ....

In the summer of 2002 lists of the top ten sustainable development successes, and the top ten failures, since Rio were published by the International Institute for Sustainable Development.[www#36] The ten successes, according to this Institute, in no particular order, were as follows:

1. Recovery of the ozone layer in our atmosphere (protects us and all life from harmful UV-B-radiation) by 2050, resulting from international co-operation to get rid of ozone-depleting chemicals such as chlorofluorocarbons and halons.
2. In direct contrast to international action, small-scale response to the Agenda 21 programme had been good, with, for example, 6000 grass-roots initiatives in over 100 countries. For example, hundreds of towns and cities (including 1300 local authorities representing more than 100 million people) had organized themselves into groupings to address specific aspects of sustainability.
3. The greater inclusion of ordinary people, interest groups and experts in the policy-making process. The rise in the number of non-governmental organizations (or NGOs) had largely fuelled this change, with 26,000 in 2002 compared with just 6000 in 1990.
4. Emergence of a greater corporate social responsibility (or jumping on the sustainability bandwagon?). Business to some extent was beginning to realize that having a social conscience, recognizing its wider responsibility towards people, communities and the environment, does not have to mean reduced profits.

5. Electronic communications, such as the internet and cell phones, were making all sorts of information more widely available and, in some ways, rendering decision-making processes more transparent.
6. Knowledge is power. Major advances had been made in the way we understand natural systems and in the development of sustainable technologies. For example, the colossal amount of cutting-edge, high-quality research in biology, chemistry, geology and physics reviewed by the Intergovernmental Panel on Climate Change was the basis of one of the most informed sets of recommendations that have ever been made to governments concerning an environmental issue. Similarly, fuel cell research has advanced significantly, taking the prospect of clean transport fuels and the so-called 'hydrogen economy' out of the realms of science fiction.
7. Measuring the progress towards (or away from) sustainability was better now owing to the development of more accurate indicators and accounting practices.
8. The Kyoto agreement on reducing greenhouse-gas emissions of 5.2% below pre-1990 levels by 2008–12. This is an interesting success story. By 2000, for many countries greenhouse-gas emissions were half of this target, mainly accounted for by a sharp fall in fossil-fuel consumption in countries of the former Eastern Bloc. The EU managed a cut of 0.5%. At the same time Japan and the US managed an increase in emissions of 3% and 13% respectively. In March 2001 the Bush administration refused to ratify the Protocol. A couple of months later at a meeting in Bonn, the Protocol was diluted, just to keep it on the rails. Kyoto made the top ten successes recognized by the Institute purely because it stayed in existence. The words 'straws' and 'clutching at' come to mind. I remember seeing on TV George Bush and Tony Blair come out of breakfast at the G8 summit in Gleneagles, Scotland (July 2005) and put Kyoto behind them with a promise of an emphasis on developing new clean technologies. These world leaders may be masters of spin but even they could not, assuming they wanted to, sell Kyoto as a success.

9. In several countries, the principles of sustainable development have begun to make inroads into the way countries are governed. However, the report acknowledges that while many politicians have learned to 'talk the talk', very few are prepared to 'walk the walk'. New Zealand is given as an example of good practice – but that's about it. Very few straws left now.
10. Environmental agreements, such as the Stockholm Convention on Persistent Organic Pollutants (signed in May 2001 by ninety-one governments and targeting the production and use of twelve specific chemicals, the so-called 'dirty dozen'), represented new protection measures for communities and the environment, although the writers of the report add the proviso that separates this final success from failure – provided that such agreements are matched by strong leadership.

... and one that didn't make it

One (partial) success story that did not make the Institute's top ten is the setting up of national parks and protected areas, as suggested by Article 8. The latest data from 2004, collated by UNEP's World Conservation Monitoring Centre, showed that almost 13% of the planet was under some kind of protection. That's up by nearly 2% since the mid-1990s. About one-quarter of western Asia, which includes countries like Turkey and the United Arab Emirates, now has protected status. In Latin America and the Caribbean it's closer to one-fifth. The greatest positive growth in protected areas has been in eastern Asia (China and Mongolia), with an increase from around 8% in the mid-1990s to almost 15% now. This stands in contrast with sub-Saharan Africa, where protected areas have hovered around 11% for the past decade.

The IUCN (World Conservation Union) estimates that there are 20,000 protected areas worldwide (land: 13.2 million square kilometres; sea: 1.3 million square kilometres), ranging from areas set aside for strict nature protection through to areas 'preserved' and managed for controlled harvesting. All of this sounds quite impressive, but it should be remembered that there are also significant problems with the present scheme of protecting areas, which takes the shine off just a little.

First, most of the areas allocated for protection are too small. True, the numbers set up increase each year, but the average size of the area is getting smaller. Setting up corridors, conduits between reserves would help, but this is rarely done.

Not only are the areas set aside too small, but they tend to be sections of land with low economic value, and rarely is it thought through in terms of the patterns of natural occurrence of the animals and plants that are found there. And while areas may be designated as national parks or wilderness areas, in few cases does this mean that they are actually afforded any protection. So in practice, many designated areas turn out not to be protected areas at all. These are referred to as paper parks. For instance, when it was set up in January 2000, the Port Honduras Marine Reserve in Belize (a marine protected area or MPA), consisting of 133 islands. There were no buoys marking the extent of the park, the large area (837 square kilometres) was difficult to patrol, and the NGO which had the responsibility for management was small with very limited resources. As a result, poaching by national and foreign fishermen went on unabated. Port Honduras Marine Reserve was a paper park. However, the next year, with the aid of a reasonable grant, the NGO built a station in the middle of the reserve which improved the ability to police the reserve and it also worked with local fishermen to develop economic alternatives to fishing, such as flyfishing, kayaking and diving and other eco-tourism developments. There are now grounds for referring to this park as a marine *protected* area. But this is an exception. Currently 80–90% of marine protected areas are not protected at all.

Finally, as recognized in the CBD but rarely acted upon, it is often not enough to set up protected areas without also protecting the wider area within which these protected zones are found. For example, there are nature reserves set aside for the African elephant. Unfortunately, four out of five elephants don't actually live in the reserves. Our commitment to operating such national parks is perhaps best illustrated by the fact that we spend a mere

US$6 million on protected areas worldwide, one-tenth of what we spend on dieting methods.

top ten failures

Given this list of 'successes', the reader might be forgiven for finding it difficult to see exactly what has changed. That there are positive signs is not in question; it is the magnitude of such signs that is hard to measure. Fortunately, the same thing cannot be said for the ten failures. As we shall see, there is a lot less 'grey area' here and many of the failures are of orders of magnitude more convincing and far-reaching than any of the successes. This is the list:

1. Failure of the Rio bargain and 'business as usual'. The most noticeable feature of the last ten years has been the conspicuous failure of the Rio 'bargain'. Collapsed leadership, failure of the developed countries to deliver and the lack of attention being focused on the institutions that had the responsibility post-Rio to co-ordinate 'on the ground' sustainable development, are all prominent.
2. The widening wealth gap. Even though global incomes are on the increase, the rich are still getting richer and the poor are still getting poorer.
3. The throw-away society is on the increase. Overconsumption is firmly entrenched as a way of life. And this is despite raised awareness and better researched information on the problems.
4. Financial aid packages with 'strings attached' have caused, and are continuing to cause, substantial damage to developing countries.
5. While ecological economics has flourished as a discipline, and although the importance of correctly valuing natural services is widely acknowledged, the environment, and therefore biodiversity, is still, literally, being sold short.
6. The world is a less peaceful place. Post cold war optimism disintegrated in the face of a growing number of 'traditional' armed conflicts, as well as a different type of war, such as terrorist attacks.

7. The oceans are still being seriously degraded owing to overfishing, coastal development and pollution.
8. Fresh water is becoming even more scarce, with knock-on effects for competition and the elevation of political tensions.
9. The tragedy of HIV/AIDS gets worse, with the greatest impact currently felt in the developing countries, especially Africa.
10. Mass species extinction continues almost unchecked.

It is difficult to exagerate this list of failures. Taken together, the ten failures and the ten successes are, at best, a reflection on our lack of commitment to these issues.

rio + 10

As I've already said, by 2002 it was widely acknowledged that progress on implementing Agenda 21, and the other action points from Rio, had been extremely disappointing, with poverty worsening and destruction of the environment continuing, and in some areas even accelerating, at alarming rates. It was recognized by the UN that what was wanted (and needed), was not philosophical or political talking shops but rather a set of concrete plans and actions, and ways of making those plans happen. In response to this need, the Commission for Sustainable Development organized a follow-up to Rio. This was the World Summit on Sustainable Development, held just outside the city of Johannesburg on 26 August–4 September 2002. With more than 22,000 people participating in one way or another, 10,000 delegates, 8000 NGOs and representatives of 'ordinary' society, 4000 media representatives, and 1000 world leaders giving addresses, this was by far the biggest international gathering ever held in Africa.[www#37]

Johannesburg had no dramatic outcome. There were no new treaties but lots of new targets, including: to halve the proportion of people without access to basic sanitation by 2015; to use and produce chemicals by 2020 in ways that do not lead to significant adverse effects on human health and the environment; to maintain or restore depleted fish stocks to levels that can produce the maximum sustainable yield on an urgent basis and where possible by 2015; and to achieve by 2010 a significant reduction in the current rate of loss of biological diversity. One of the key differences between Johannesburg and Rio was the fact that different approaches were taken to the problems raised. The 2002 meeting involved far more dialogue with the groups originally identified in Agenda 21 – business, farmers, indigenous peoples, industry, local authorities, NGOs, scientists, technologists, trade-unionists, 'women', 'workers' and young people. And commitments were made not just by governments but also by a host of others – NGOs, intergovernmental organizations and by businesses. More than 300 voluntary initiatives were launched.

US Secretary of State Colin Powell declared the Summit a 'successful effort'. However, not all the participants were equally as pleased. Venezuelan President Hugo Chavez (the man who was apparently selling petrol at twelve cents a gallon to Venezuelans – but hey ho), representing the views of 132 developing countries, while agreeing it was successful, was equally clear that it was still not enough, and in fact some of the steps taken could have been considered as steps backward. Many of the NGOs also thought that the summit did not go far enough, particularly when it came to target setting for increasing the use and development of renewable energies. While disappointed with some of the outcomes, the President of the World Resources Institute, Jonathan Lash, commented positively that: 'This Summit will be remembered not for the treaties, the commitments, or the declarations it produced, but for the first stirrings of a new way of governing the global commons – the beginnings of a shift from the stiff formal waltz of traditional diplomacy to the jazzier dance of improvisational, solution-oriented partnerships that may include non-government organizations, willing governments and other stakeholders.'

millennium assessment

To start the first year of a new century and a new millennium on a good footing, the UN produced Resolution 55/2, The United

Nations Millennium Declaration, which contained eight specific goals:[www#38] to eradicate extreme poverty and hunger; to achieve universal primary education; to promote gender equality and empower women; to reduce child mortality; to improve maternal health; to combat HIV/AIDS, malaria and other diseases; to ensure environmental sustainability; and to develop a global partnership for development. They are referred to collectively as the UN Millennium Development Goals. All 191 UN member states have pledged to meet these goals by 2015. The resolution contained a section on respect for nature, which says that: 'The current unsustainable patterns of production and consumption must be changed in the interest of our future welfare and that of our descendants.' It also specifically states that the ongoing and worsening degradation of ecosystem services is a stumbling block to these Millennium Development Goals.

Against this backdrop, in 2001 the UN initiated a massive study which brought together more than one thousand experts, representing ninety-five different countries, to: (a) produce a global inventory of the state of our ecosystems; (b) quantify the effect that human activities are having on them; and (c) make suggestions for the future. The results were published in the summer of 2005 under the title *The Millennium Ecosystem Assessment Synthesis Report.*[www#26] It states that: 'Any progress achieved in addressing the goals of poverty and hunger eradication, improved health, and environmental protection is unlikely to be sustained if most of the ecosystem services on which humanity relies continue to be degraded.' Comprehensive in both its coverage and depth, one of the Report's key outcomes is that continuing degradation of fifteen of the twenty-four ecosystem services that the earth provides is increasing the likelihood of abrupt changes that will have a serious and negative impact on our well-being. It gives examples of such changes – sudden changes in water quality, shifts in climate, fisheries collapse and the emergence of new diseases. At the time of writing, it is too early to gauge any response to the Millennium Assessment.

no room for the individual?

It is widely accepted that individuals can have a great influence on conserving biodiversity and the environment at the local level. But the global problems, as we've seen, seem insurmountable even for governments and intergovernmental committees. What possible influence could an individual have on a global scale?

Wangari Muta Maathai was born in Nyeri, Kenya in 1940. She was the first woman in east and central Africa to earn a doctorate degree. She became Chair of the Department of Veterinary Anatomy at Nairobi in 1976, the first woman to hold such a position in the region. While there, she was very active in the National Council of Women of Kenya between 1976 and 1987, being its chairperson from 1981 onwards. It was during this time that she hit upon the idea of planting trees with women's groups in order both to conserve the environment and to improve their quality of life. During her speech accepting the Nobel Peace Prize in 2004, she said, 'My inspiration partly comes from my childhood experiences and observations of nature in rural Kenya ... As I was growing up, I witnessed forests being cleared and replaced by commercial plantations, which destroyed local biodiversity and the capacity of the forests to conserve water.'[www#39]

Through the organization she formed, the Green Belt Movement, she has assisted women in planting more than 20 million trees on their farms and in school grounds and in church compounds. The organization expanded into the Pan African Green Belt Network in 1986 and similar tree-planting schemes were extended to countries such as Ethiopia, Lesotho, Malawi, Tanzania, Uganda and Zimbabwe. She played a major role in seeking the cancellation of African debt, through the Jubilee 2000 Coalition, and her hands-on protests against land grabbing and deforestation have earned her a high international profile. By 2000 she was internationally recognized for her campaigning for democracy, human rights and environmental conservation. In December 2002, Professor Maathai was elected to parliament with an overwhelming 98% of the vote and was appointed

Assistant Minister for Environment, Natural Resources and Wildlife in Kenya's parliament.

In October 2004 she was awarded the Nobel Peace Prize 'for her contribution to sustainable development, democracy and peace'. Not everyone was pleased with the choice. One politician raged, 'You don't give the Nobel chemistry prize to a professor in economics. A peace prize should honour peace, not the environment.' Also her controversial (and, if rightly reported, wrong) views on the origin and spread of AIDS have caused a bit of a stir. However, in many ways her nomination underlines the point made in this and the previous chapter – that you cannot really separate concern for biodiversity from concern for the environment generally, and from the multitude of other social and political problems, including poverty, that seem to go hand in hand. This can even be seen in her acceptance speech, where she said, 'So, together, we have planted over 30 million trees that provide fuel, food, shelter, and income to support their children's education and household needs. The activity also creates employment and improves soils and watersheds.'

Both Wangari and her Green Belt Movement have received numerous other awards. She has been listed on UNEP's Global 500 Hall of Fame and named one of the hundred heroines of the world. In June 1997 she was elected by *Earth Times* one of 100 persons in the world who have made a difference in environmental conservation.

Wangari concluded her Nobel Prize acceptance speech with these words:

> I reflect on my childhood experience when I would visit a stream next to our home to fetch water for my mother. I would drink water straight from the stream. Playing among the arrowroot leaves I tried in vain to pick up the strands of frogs' eggs, believing they were beads. But every time I put my little fingers under them they would break. Later, I saw thousands of tadpoles: black, energetic and wriggling through the clear water against the background of the brown earth. This is the world I inherited from my parents.

> Today, over 50 years later, the stream has dried up, women walk long distances for water, which is not always clean, and children will never know what they have lost. The challenge is to restore the home of the tadpoles and give back to our children a world of beauty and wonder.

Can one person make a difference? From Wangari's story the answer must be yes. She's an exceptional person, admittedly, but one person nevertheless – and that person doesn't have to be right in everything they believe.

epilogue

There have been great advances in our thinking about what to conserve and how to conserve, even on a global scale. The end of the last century saw the first global attempt to tackle our global biodiversity crisis, given substance in the words of the Convention for Biological Diversity. Unfortunately, as we have seen, our actions and resolve have not kept pace with either our enlightened thinking or our words. The political and moral will is still lacking to take the conservation of biodiversity seriously. Despite mounting scientific evidence, much of it presented here in previous chapters, there is still an unwillingness to face the problem of our biodiversity crisis. One academic wrote, 'there is unfortunately no precedent for 5 billion human beings suddenly sharing an enlightened vision of the future'.

This morning I forgot to sort out the plastic, glass and paper for the different recycling bags our city council supplies us with. I heard the refuse collectors at the top of our street, So I stuffed all the rubbish together into a number of black rubbish bags for collection, just pleased that I hadn't missed it entirely and so would not have rubbish lying around for another week. And that was the action of someone who says he knows how bad our biodiversity crisis is, having studied it and even written about it.

chapter eight

the ring of living beauty

This is to me the most beautiful and saddest landscape in the world.

The Little Prince, Antoine de Saint Exupéry

It was three o'clock in the afternoon, with an hour and a half to go before low tide. Such a beautiful summer's day. The water in the rock pools was warm to the touch, almost glistening and sparkling in the sun's radiant light. My children, Ellie, Ethan and Ben, clambered carefully over the rocks of the intertidal zone, with me following close behind. Ben and Ethan found a particularly beautiful and rich pool. Carefully moving the brown and red seaweeds to one side, they methodically inspected the underside of a large flat rock for marine invertebrates. Within three or four minutes we'd seen twenty or more species of relatively large animals. The underside of the rock was peppered with tiny, circular spirobid worms, sheeted with two different types of sea mat, three different-coloured breadcrumb sponges and an incredible colony of star sea squirts. Some small amphipod crustaceans sideways-scuttled off the rock, followed by numerous hairy and naked porcelain crabs dropping like stones back into the water. Three purple-tipped sea urchins clung with their assorted shell bits and weed hats to the far corner of the stone, almost eclipsing two closely huddled sea slugs.

Each new find was greeted with a 'Wow!', or a gasp, or a 'Look at that!' Three children, their attention totally captivated by the wealth, beauty and awe of so many different living things, searched and watched and thought. By this time some other children, with associated parents, had arrived, and I heard my children tell them with an infectious enthusiasm about what they'd found. You'd think they'd get fed up with all this rock-pooling stuff. My kids aren't really 'into wildlife' as such and I've been dragging them along shores since they were born. But it doesn't seem to matter. The pull is strong.

I heard one of the parents exclaim, 'I never knew all this stuff existed here.'

'Wembury's famous for its marine life,' I said.

All at once Ethan picked up a squat lobster that had just emerged from some weeds. Everyone let out screams of surprise and even joy.

'Wow, look at that!' 'That's so cool.' 'Can I hold it?'

'Will this stuff always be here?' asked one of the mums.

Before I could answer, Ethan replied, 'It'll probably be extinct soon.'

'No, I mean tomorrow,' the woman pointed out.

We all laughed, but deep down I felt ashamed. The truth of Ethan's words cut deep. They brought to mind the book *Father and Son* by Edmund Gosse. After describing the beauty of the Devonian rock pools he saw with his father Phillip, Edmund reminisces sadly: 'All this is long gone, and done with. The ring of living beauty drawn about our shores was a very thin and fragile one ... (n)o one will see again the shore of England that I saw in my early childhood.'

When Kevin Gaston and I were working on our textbook *Biodiversity*, I remember one evening reading all the latest scientific papers on calculating extinction rates, and checking out how rigorous the information on known extinctions was. I found myself putting down my notes and papers and going quietly into each of my children's rooms. I watched Ellie, Ethan and Ben sleeping peacefully and found a deep sadness within me. I find that tears come to me more readily as I grow older. At best, what they would inherit from me would be a world considerably poorer in terms of the diversity of living things; the ring of living beauty (remember the football in the garden?) was a thin and fragile one. That sadness has stayed with me since. It is one of the things that drove me to write this book.

We are in the midst of the biodiversity crisis, outstripping anything we've seen in the past. In books and articles I read in the 1970s, and even in some works still written today, there is talk of how to avert the crisis. This is not a current option. It's here – present tense. True, it is difficult to get a good scientific grip on the science of biodiversity, and there are still, as we've seen time and time again, many areas where our information is incomplete and our understanding poor. But even the little we do know is enough to convince us of the seriousness of our situation. And still many of us play games with what came out of Rio and Johannesburg. I do not think, as the real scaremongers preach, that everything will go extinct and life will be extinguished from our planet. It is a possibility, but I think unlikely, at least not yet. The truth, I believe, is much worse. The question is not how can we avert our biodiversity crisis, but what sort of world do we want to live in.

Rachel Carson, an eminent marine biologist from the first half of the twentieth century, wrote an influential book called *Silent Spring.*[www#40] In it she pictured a world devastated by the indiscriminate use of pesticides like DDT. The world she wrote of was a world devoid of birds singing. The image was so powerful, and the book so moving, that it changed the way a whole generation thought about pollution. Faced with the question 'What sort of world do you want to live in?', the readers of Carson's book, and the world in general, opted for one where birds still sang.

So, for those of us living now at the beginning of this new millennium, the question is the same – 'What sort of world do you want to live in?' Do you want to live in a world where, increasingly, economic and physical wars are fought over water and energy and access to other resources, where starvation and poverty are even more prevalent because the people most at risk have destroyed much of the biodiversity that gave them the ability to exist, never mind live? A world where only the rich can pay the enormous amounts of money to secure dwindling services that once were free? A world devoid, stripped of all natural beauty, where the variety of life is something that only exists in books or on old photo CD ROMS? A world where the predominant 'natural' colour is grey and where mere existence, just getting by, just hanging on a little longer, becomes an even more acceptable and inevitable option over life? A world where, when you take your children to what remains of the beach, you spend hours searching for just a few measly crabs and worms, as the great diversity you once saw and loved as a child has long gone?

An interest in biodiversity is not just 'a passion for wildlife' or a 'commitment to favourite conservation issues', although it may encompass those things. It's not just a scientific discipline, although it can be studied as such – as we have done – and much pleasure can be derived from that study alone. Biodiversity is intimately and inextricably tied up with our existence, our lives and our lifestyles. It doesn't matter whether you've got green tendencies, or you're a hard-nosed business person, whether you think that saving the lesser spotted weevil from extinction should be an all-consuming passion, or your days are spent merely getting by to make sure your family is OK. The bottom line is, no matter who you are, biodiversity matters. And it matters to you, your family, everyone you know, and everyone you do not know and will never meet, both at the highest levels and at the most fundamental levels of existence. And that's why it's worth knowing about. We cannot avert our biodiversity crisis. But we should not allow the question 'What sort of world do we want to live in?' to go unanswered or even unconsidered. As pointed out by Mother Teresa towards the end of the twentieth century, poverty is so much more than not having money. It is a dreadful thing to waste our lives, and the lives of all the living things around us, by default.

For those who, like me, have belief in a God who created all this biodiversity and to whom biodiversity belongs, the realization that we have been bad stewards should fill us with sadness. My understanding is that while environmental concerns do not seem to be high up the priority list of many of the Christians with whom I share my faith, it does rank high in the purposes of our God. And this demands some response – 'What sort of world does God want us to live in?' I know it is not the one I am passing on to my children.

By this time some of you will have switched off, either in disbelief (on so many different levels) or because the picture I've painted is too terrible even to contemplate. But for others there will be a feeling that says, 'OK, you've pointed out the problems, now present us with some options.' What should we do? I believe the first thing we must do in the midst of this biodiversity crisis is actually believe that we as individuals can make a difference. I love the film *One Flew over the Cuckoo's Nest*, with Jack Nicholson. There's a scene in which he tries to escape from the psychiatric hospital where he finds himself. In the wash area, and in front of a group of other patients, he proclaims that he's going to lift up this huge water font and throw it through the wall, so making good his escape to the outside world. Nicholson struggles with moving, let alone lifting, the water font and finally, exhausted, he has to give up. The surrounding patients give off an air of 'we knew it couldn't be done' and Nicholson shouts angrily at them, 'At least I tried, goddammit, at least I tried.' Stirring stuff. The film ends with one of the patients, a Native American Indian, 'The Chief', returning to the washroom and, after an incredible effort, picking up the water font and throwing it through the wall, making good his escape.

We can lobby our politicians, we can get involved in local politics, we can think of innovative ways of conserving biodiversity, locally, nationally, even globally. We can change the way we live, but most of all we must believe that we can change things even in the midst of our biodiversity crisis. As someone has said, if we, the generation that faces this next century, does not do the impossible, we will be faced with the unthinkable. And I do not see why my children, our children, should be faced with that.

further reading and website references

If what you've read has whet your appetite for more, some books I would recommend are listed below. This list is nowhere near exhaustive. It is just a start – for beginners. Certainly, the textbooks I referred you to at the end of chapter 1 are the first port of call for those who want to tackle seriously the academic discipline which is biodiversity. The website addresses listed here are referenced in the text and they too will help you go much further and deeper into understanding and conserving biodiversity.

textbooks

Gaston, K. J. and Spicer, J. I. 2004. *Biodiversity. An Introduction* (2nd edition). Blackwell.

Jeffries, M. J. 1997. *Biodiversity and Conservation.* Routledge.

Lévêque, C. and Mounolou, J. C. 2001. *Biodiversity.* John Wiley.

Meffe, G. K., Carroll, C. R. and Groom, M. J. 2004. *Principles of Conservation Biology.* Sinauer.

authoritative reference works

Groombridge, B. and Jenkins M. D. 2002. *World Atlas of Biodiversity. Earth's living resources in the 21st century.* University of California Press.

Heywood, V. H. (ed.) 1995. *Global Biodiversity Assessment.* Cambridge University Press.

Levin, S. A. (ed.) 2001. *Encyclopedia of Biodiversity.* 5 volumes, Academic Press.

Millennium Ecosystem Assessment 2005. *Ecosystems and Human Well-being: Biodiversity Synthesis.* World Resources Institute. (Can be purchased as a hard copy or downloaded in PDF format from http://www.maweb.org/en/products.aspx)

popular books

Adams, W. M. 1997. *Future Nature. A Vision for Conservation.* Earthscan.

Barbier, E. B., Burgess, J. C. and Folke, C. 1994. *Paradise Lost? The ecological economics of biodiversity.* Earthscan.

Beattie, A. and Ehrlich, P. 2001. *Wild Solutions: How biodiversity is money in the bank.* Yale University Press.

Berry, R. J. 2000. *The Care of Creation. Focusing concern and action.* Inter-Varsity Press.

Bourne, J. and Jones, E. 2003. *Go M.A.D.! Go Make a Difference: Over 500 daily ways to save the planet.* Think Publishing.

Carson, R. 2000. *Silent Spring.* Penguin Classics.

Drury, S. 2001. *Stepping Stones: The makings of our home world.* Oxford University Press.

Flannery, T. and Schouten P. 2001. *A Gap in Nature. Discovering the world's extinct animals.* William Heinemann.

Fuller, E. 2002. *Dodo: From Extinction to Icon.* Collins.

Leakey, R. and Lewin, R. 1996. *The Sixth Extinction: Biodiversity and its survival.* Phoenix.

Lovelock, J. 2000. *Gaia: A new look at life on Earth* (3rd edition). Oxford University Press.

Mackay, R. 2002. *The Atlas of Endangered Species.* Earthscan.

Margulis, L., Schwartz, K. V., Dolan, M. and Delisle, K. 1999. *Diversity of Life on Earth: The illustrated guide to the five kingdoms.* Jones and Bartlett Publishers International.

Margulis, L. and Schwartz, K. V. 1998. *Five Kingdoms: An illustrated guide to the phyla of life on earth* (3rd edition) W. H. Freeman.

Maslin, M. 2004. *Global Warming: A very short introduction.* Oxford University Press.

McKibben, B. 2003. *The End of Nature.* Bloomsbury.
McNeill, J. 2001. *Something New under the Sun: An environmental history of the world in the 20th century.* Penguin.
Pimm, S. L. 2001. *The World According to Pimm: A scientist audits the earth.* McGraw-Hill.
Safina, C. 1997. *Song for the Blue Ocean: Encounters along the world's coasts and beneath the seas.* Henry Holt.
Shepherd, A. and Oakley, C. 2004. *52 Weeks to Change Your World.* Centre for Alternative Technology.
Stone, D., Ringwood, K. and Vorhies, F. 1997. *Business and Biodiversity: A guide for the private sector.* IUCN.
Suzuki, D. 1999. *The Sacred Balance: Rediscovering our place in nature.* Bantham.
Terborgh, J. 1999. *Requiem for Nature.* Island Press.
Tudge, C. 2002. *The Variety of Life. A survey and celebration of all the creatures that have ever lived.* Oxford University Press.
Wilcove, D. S. 2000. *The Condor's Shadow: The loss and recovery of wildlife in America.* Anchor Books.
Wilson, E. O. 2001. *The Diversity of Life.* Penguin.
Wilson, E. O. 2002. *The Future of Life.* Little Brown.
Worldwatch Institute 2006. *State of the World 2006.* WW Norton & Company.

website references

1. The full text of the Rio document can be found at *http://www.biodiv.org/convention/default.shtml*, together with notes on what has happened since.
2. Type 'Wembury' into *http://www.goodbeachguide.co.uk/* and you'll get an idea of the place.
3. The full text of the *Plymouth marine fauna* can be found online at *http://www.mba.ac.uk/PMF/*.
4. To order a copy of our book, log on to Blackwell online publishing at *http://bookshop.blackwell.co.uk/*. You can also get the Lévêque and Mounolou book from the same site.
5. For an introduction to just about any group that I refer to and many, many more besides use the taxon-lift at *http://www.ucmp.berkeley.edu/help/taxaform.html*. I think this is an excellent online

resource for exploring the different designs that make up life on earth. The Tree of Life project at *http://tolweb.org/tree/phylogeny.html* also provides easy access to information about the diversity of organisms on earth.

6. Introduction to arthropods at *http://www.ucmp.berkeley.edu/arthropoda/arthropoda.html.*
7. Introduction to the mushrooms and moulds at *http://www.ucmp.berkeley.edu/fungi/fungi.html* but also check out *http://www.doctorfungus.org/* for a very accessible site which provides an excellent introduction to fungi for the non-specialist.
8. Introduction to bacteria at *http://www.ucmp.berkeley.edu/bacteria/bacteria.html* but if you're like me, you may well benefit more from the children's introduction to microbes found at *http://www.microbe.org/*. I also thought that the Microbe Zoo from the Digital Learning Centre for Microbial Ecology was an entertaining but authoritative introduction to microbes: *http://commtechlab.msu.edu/sites/dlc-me/zoo/index.html.*
9. Go to this site *http://www.ucmp.berkeley.edu/alllife/eukaryotasy.html*, where you can choose to explore any of the protoctistan groups.
10. There are good introductions to nematodes and parasitic nematodes in particular at *http://www.ucmp.berkeley.edu/phyla/ecdysozoa/nematoda.html* and *http://nematode.unl.edu/* respectively.
11. See *http://www.virology.net/Big_Virology/BVHomePage.html* for the big picture book of viruses and *http://www.ucmp.berkeley.edu/alllife/virus.html* for a basic introduction.
12. Introduction to plants, group by group, can be found at *http://www.ucmp.berkeley.edu/help/index/plantae.html.*
13. An introduction to molluscs is at *http://www.ucmp.berkeley.edu/mollusca/mollusca.html.*
14. An introduction to the animals with a backbone, the vertebrates, can be found at *http://www.ucmp.berkeley.edu/vertebrates/vertintro.html.*
15. Just in case you are interested in knowing more about this excellent course, you can find out at *http://www.plymouth.ac.uk/courses/course.asp?id=0732.*
16. Lots of good stuff on biodiversity hotspots from Conservation International can be found at *http://www.biodiversityhotspots.org/xp/Hotspots.* Much of it is an elaboration of Myers' original scheme.
17. Here are two sites which attempt to put together an Atlas of Biodiversity. At *http://stort.unep-wcmc.org/imaps/gb2002/book/*

viewer.htm there is an interactive world atlas of biodiversity and at *http://www.nhm.ac.uk/research-curation/projects/worldmap/index. html* you can find a site put together by the Natural History Museum in London which, through a special project, World Map, seeks to map biodiversity globally and identify priority areas for conservation.

18. For a profile of E. O. Wilson go to *http://www.metrokc.gov/dnrp/swd/naturalconnections/edward_wilson_bio.htm*
19. Here is a site (*http://www.ucmp.berkeley.edu/vendian/critters.html*) about Ediacaran (sometimes called Vendian) animals, with some really nice pictures of these bizarre creatures.
20. *http://www.burgess-shale.bc.ca/* is the official Burgess Shale website from the Burgess Shale Geoscience Foundation, but there is also a good introduction at *http://www.ucmp.berkeley.edu/cambrian/burgess.html.*
21. There is an excellent website by the University of Aberdeen on the Rhynie Chert *http://www.abdn.ac.uk/rhynie/*, where you can also check out the first land arthropods (*http://www.abdn.ac.uk/rhynie/faunbasic.htm*).
22. Check out the BBC website on mass extinctions at *http://www.bbc.co.uk/education/darwin/exfiles/massintro.htm.*
23. You can find a comprehensive source of information on the current mass extinction at *http://www.well.com/user/davidu/extinction.html* and an excellent online article by Niles Eldredge entitled 'The sixth Extinction' at *http://www.actionbioscience.org/newfrontiers/eldredge2.html.*
24. The IUCN Red Data Book for 2004 is online at *http://www.redlist.org/*. For more detail on different vulnerable groups and conservation check out the links page at *http://www.redlist.org/info/links.*
25. You can download the actual Living Planet Reports from *http://www.panda.org/news_facts/publications/key_publications/living_planet_report/index.cfm.*
26. Official site of the Millennium Ecosystem Assessment can be found at *http://www.maweb.org/en/index.aspx.*
27. *http://www.ipcc.ch/* is probably the best, most authoritative site for climate change information, hosted by the Intergovernmental Panel on Climate Change, although the UN framework convention on climate change is also worth a look *http://unfccc.int/2860.php.*
28. *http://www.ecotourism.org/* is a US-based international ecotourism society set up for researchers, conservationists and business. The UNEP ecotourism website at *http://www.unep.fr/pc/tourism/*

ecotourism/home.htm is also useful both in its own right and because of an excellent 'links' page.

29. The population reference bureau informs people around the world about population, health and the environment, and empowers them to use that information to advance the well-being of current and future generations. Their website is at *http://www.prb.org/*. If you go to *http://www.ibiblio.org/lunarbin/worldpop* you will find a counter which gives you in real time the number of people on earth or if you prefer the number of people at any time in the past.
30. James Lovelock's homepage can be found at *http://www.ecolo.org/lovelock/index.htm*.
31. The official website of the Biosphere 2 project can be found at *http://www.bio2.com/*. But it's also worth checking out *http://www.bioquest.org/simbio2.html*, an interactive site which models the Biosphere 2 project and allows you to re-run the experiment, changing things like the number of people in the closed system and what sort of organisms you would include.
32. In terms of the sorts of organizations that are and have been involved in conservation in its widest sense, it's worth checking the websites for the UN Environment Programme, *http://www.unep.org/*, as well as the joint venture with the World Conservation Monitoring Centre, *http://www.unep-wcmc.org/*. There is also the WWF, *http://www.wwf.org/*, and the Sustainable Development Department of the Food and Agriculture Organization (FAO for short) of the United Nations, *http://www.fao.org/WAICENT/FAOINFO/SUSTDEV/index_en.htm*.
33. The whole of this landmark book can be read online at *http://books.nap.edu/books/0309037395/html/index.html*.
34. The full text of the UK Biodiversity Action Plan can be found at *http://www.ukbap.org.uk/*.
35. Website of the UN Department of Economic and Social Affairs, Division of Sustainable Development is found at *http://www.un.org/esa/sustdev/index.html*.
36. Visit *http://www.iisd.org/briefcase/ten+ten.asp* and read about the top ten successes and failures in much more detail than I've presented here.
37. Official website of the Johannesburg World Summit on Sustainable Development 2002 can be found at *http://www.johannesburgsummit.org/*. The official site of the 2002 earth summit can be found at *http://www.earthsummit2002.org/*.

38. You can find the UN Millennium Development Goals in full at *http://www.un.org/millenniumgoals/* and an official website that gives a list of indicators to check the extent to which these goals are being achieved *http://millenniumindicators.un.org/unsd/mi/mi_goals.asp*.
39. For more details on Wangari Muta Maathai, see her pages at the Nobel prize website *http://nobelprize.org/peace/laureates/2004/*.
40. A website dedicated to Rachel Carson and her work is at *http://www.rachelcarson.org/*.

appendix

classification of the groups of organisms referred to in this book

DOMAIN[#1] EUKARYOTA

KINGDOM [#1] ANIMALIA

Phylum Annelida – true or segmented worms
Phylum Arthropoda – jointed-leg, segmented animals
  Subphylum Chelicerata – mites, scorpions, spiders and ticks
  Subphylum Crustacea – barnacles, crabs, lobsters, shrimps, woodlice
  Subphylum Trilobita – extinct
  Subphylum Uniramia – centipedes, millipedes and insects
Phylum Brachiopoda – lampshells
Phylum Bryozoa – moss animals
Phylum Chordata
  Subphylum Cephalochordata – lancelets
  Subphylum Urochordata – tunicates and sea squirts
  Subphylum Vertebrata
    Class Agnatha – jawless fish such as lampreys and hagfish
    Class Amphibia – frogs, toads and salamanders
    Class Aves – the birds
    Class Chondrichthyes – cartilaginous fish such as sharks, rays, skates

Class Mammalia – warm-blooded, fur-covered creatures that suckle their young, including humankind as a biological entity

Class Osteichthyes – bony fish. A mixed bag including lobe-finned, e.g. lungfish and coelacanth, and ray-finned fish, e.g. sturgeon, salmon, seahorses, perch

Class Reptilia – snakes, lizards and turtles

Phylum Cnidaria – radial symmetry with stinging cells called nematocysts

Class Hydrozoa – 'plant' animals

Class Scyphozoa – jellyfish

Class Anthozoa – sea anemones and corals

Phylum Echinodermata

Class Asteroidea – starfish

Class Crinoidea – sea lilies or featherstars

Class Echinoidea – sea urchins and sand dollars

Class Holothuroidea – sea cucumbers

Phylum Hemichordata – acorn worms and the nearest living relatives to the extinct graptolites

Phylum Mollusca

Class Polyplacophora – chitons

Class Gastropoda – snails, sea slugs and land slugs

Class Bivalvia – clams, mussels, oysters, scallops

Class Cephalopoda – squid, octopus, cuttlefish, extinct ammonites

Phylum Nematoda – roundworms

Phylum Nemertea – ribbon or bootlace worms

Phylum Onychophora – velvet worms

Phylum Placozoa – simplest known animal

Phylum Platyhelminthes – flatworms, but includes some important parasites such as tapeworms

Phylum Porifera – sponges

Phylum Tardigrada – the water 'bears'

KINGDOM #2 FUNGI

Phylum Zygomycota – includes the breadmoulds

Phylum Ascomycota – most of the yeasts, moulds and truffles
Phylum Basidiomycota – club fungi, the most diverse group
[LICHENS – don't quite fit as they consist of an interaction between Fungi and bacteria and/or green algae]

KINGDOM #3 PLANTAE

Phylum Anthophyta – angiosperms or flowering plants
'Phylum' Bryophyta – mixed bag that probably should be split into a number of phyla, comprising bryophytes, liverworts and mosses
Phylum Coniferophyta – Conifers
Phylum Cycadophyta – cycads
Phylum Ginkgophyta – Ginko or maidenhair tree
Phylum Lycophyta – lycophytes and club mosses
Phylum Psilophyta – whisk ferns
Phylum Pteridospermophyta – seed ferns
Phylum Pterophyta – ferns
Phylum Sphenophyta – horsetails

KINGDOM #4 PROTISTA OR PROTOCTISTA

Phylum Acrasiomycota & Myxomycota – slime moulds
Phylum Apicomplexa – totally parasitic protozoans
Phylum Charophyta – stoneworts
Phylum Chlorophyta – green seaweeds
Phylum Chrysophyta – small alga including diatoms
Phylum Chytridiomycota – protozoans but closely related to the Fungi
Phylum Ciliophora – ciliated protozoan
Phylum Euglenophyta – small 'single-celled' creatures that can alternate between animal-like and plantlike features
Phylum Sarcodina – amoebalike protozoans, sometimes encapsulated in beautifully sculptured, if tiny, shells
Phylum Mastigophora – mixed bag of small single-celled creatures, many with a tail-like structure (a flagellum) used in locomotion

Phylum Oomycota – water moulds
Phylum Phaeophyta – brown seaweeds
Phylum Rhodophyta – red seaweeds
Phylum Pyrrhophyta – dinoflagellates

KINGDOM #5 MONERA

Microbial life
(In the three domains scheme this fifth Kingdom is split between the remaining two DOMAINS, ARCHAEBACTERIA and EUBACTERIA.

index

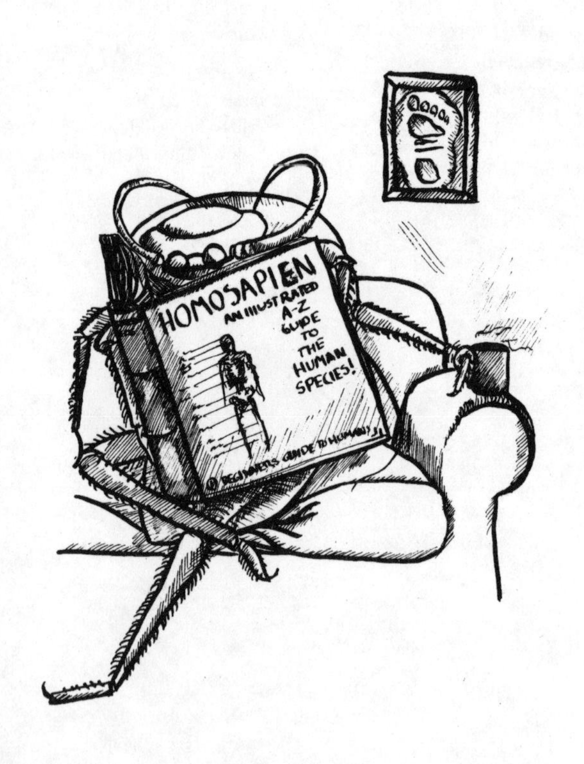
HOMOSAPIEN
AN ILLUSTRATED A-Z GUIDE TO THE HUMAN SPECIES!